PUBLISHERS' NOTE

The series of monographs in which this title appears was introduced by the publishers in 1957, under the General Editorship of Dr Maurice G. Kendall. Since that date, more than twenty volumes have been issued, and in 1966 the Editorship passed to Alan Stuart, D.Sc.(Econ.), Professor of Statistics, University of London.

The Series fills the need for a form of publication at moderate cost which will make accessible to a group of readers specialized studies in statistics or courses on particular statistical topics. Often, a monograph on some newly developed field would be very useful, but the subject has not reached the stage where a comprehensive treatment is possible. Considerable attention has been given to the problem of producing these books speedily and economically.

It is intended that in future the Series will include works on applications of statistics in special fields of interest, as well as theoretical studies. The publishers will be interested in approaches from any authors who have work of importance suitable for the Series.

CHARLES GRIFFIN & CO. LTD

GRIFFIN'S STATISTICAL MONOGRAPHS AND COURSES

For a list of other statistical and mathematical books
see inside of back cover.

CUMULATIVE SUM TESTS
THEORY AND PRACTICE

C. S. VAN DOBBEN DE BRUYN

Operations Research Manager, Mars Chocoladefabriek N.V.
Veghel, Holland
(formerly with Philips Industries Ltd, Croydon, Surrey)

BEING NUMBER TWENTY-FOUR OF
GRIFFIN'S STATISTICAL MONOGRAPHS & COURSES

EDITED BY
ALAN STUART, D.Sc.(Econ.)

1968
HAFNER PUBLISHING COMPANY
NEW YORK

TO RIA

PREFACE

Control charts were originally invented to monitor the quality of manufactured products. There has been a gradual development of these quality-control procedures from 1931 onwards – a development which was accelerated considerably in the late 'fifties by research into cumulative sum tests ("cusum tests"). Tests based on the cumulative sum of deviations from target have proved to be generally sharper than any of the tests in existence previously: a very similar, and indeed related, success was scored some years earlier by sequential tests for acceptance sampling.

The object of this monograph is to provide a practical guide to the use of cusum tests in a variety of situations which occur in industrial practice; moreover, their properties are investigated using suitable criteria. Much of the material presented in the monograph has been published elsewhere, and the author has perused the articles listed in the bibliography during the progress of his own work. The terminology chosen is an attempt to use the most common terms from the existing literature.

Works in the bibliography are referred to by the name of the author(s) followed by the year of publication, e.g. Barnard (1959).

The book consists of three main parts:

(1) An introduction, in which cusum tests are introduced and placed in their proper context. Definitions of the terms used are also given in this first part.

(2) A guide to practical applications, with tables. Most of the tables have been computed by the author on a large digital computer (Control Data 3600); they appear here for the first time.

(3) Finally a more theoretical part follows, in which analytic and numerical ways of solving the integral equations for Average Run Length and other quantities are presented.

During the later stages of preparation of the manuscript two papers were published, which are highly relevant to the subject-matter of the monograph: these are Kemp (1967a) and Kemp (1967b) as given in the bibliography. The inequalities given in Kemp (1967b) impose stricter limits than those in section 1.6 of this monograph, and they are of practical value in choosing a set of test parameters.

I am indebted to Philips Industries Ltd. of Croydon for encouraging me to study the subject in great detail. I am even more indebted to all those persons, both at Philips Industries and elsewhere, who have at one time or another, perhaps unknowingly, contributed to the birth of this monograph. One person I should like to mention by name for his unusually great contribution: my friend A.R.W. Muijen in Eindhoven (Holland), who always sorted out the occasional misunderstandings between me and the Control Data 3600 computer.

February 1968   C.S. VAN DOBBEN DE BRUYN

CONTENTS

1 INTRODUCTION

1.1 General introduction

Let the process under scrutiny produce observations in numerical form x_1, x_2, x_3, ... , generated in this order. These represent observations of the level of some quantity relevant to the state of the process to be controlled. They could be the diameters of spindles produced successively on an automatic lathe, or the number of packets of soap sold per week in a particular area: and indeed, any quantity that needs control of its level or only an estimate of its level.

Also associated with the process there is a "target value" X. The process is said to be "in control" if the mean value of the measured quantity is close to the target value. In the case of the spindles from the automatic lathe, X is the specification diameter of the products. In the case of the packets of soap the target value may be a sales objective, which is to be attained by the right amount of promotion, or it may just be the sales forecast. In the latter case the sales level is not being scrutinized with the purpose of controlling promotion, but rather with a view to reporting of inaccuracies in the forecast.

Let us define what is exactly meant by the expressions: "the process is in control" and "the process is out of control". Most processes exhibit a certain amount of inherent variability. Differences between the observations and the target value will always occur in that case. It is important to distinguish clearly between random variations and systematic errors due to the process being out of control. On p.16 of Davies (1963) the distinction between *accuracy* and *precision* is lucidly explained. The analogue of the example used there for the case of the automatic lathe is the following.

The lathe may be a new one, of excellent quality, so that it will produce spindles with diameters very close to one another: the natural variation of the process is small in that case. We call this a lathe which produces spindles with great precision. However, somebody may have made a mistake in adjusting the machine settings, taking the calibrated inch marks for centimetre marks. The machine is then producing spindles of which the diameters are *with great precision* too large by a factor of $2{\cdot}54$; the process has a *very low accuracy* with respect to the target, when this mode of machine setting is adopted. The process is indeed badly out of control, making products which deviate systematically from the specification.

Conversely, as the lathe becomes older, its precision will proba-

2

bly decrease because of wear, so that the natural variability increases; the accuracy, on the other hand, will probably have improved, because the operators have got used to the peculiarities of the machine and have learned how to set it.

Strictly speaking a process is only in control when the maximum possible accuracy is achieved; in other words, when no systematic bias exists at all. All other situations can be classified as out-of-control situations. In practice, however, this is not usually attainable or even desirable. The process is considered to be satisfactorily in control if the bias is less than a specified, fairly small, percentage of the standard deviation of the x's. This standard deviation is a measure of the "lack of precision" of the process. The bias is defined as the difference between the true (population) process mean and the target value; in the sequel we shall always simply refer to "bias".

Control charts serve to distinguish between "in-control" and "out-of-control" situations. If there were no inherent variability of the process, there would be no need for more or less sophisticated control charts, because the slightest difference between a value x_i and the target value X would indicate lack of control and appropriate action could be taken. It is just because the systematic bias is embedded in "noise" (a now popular term to describe random fluctuations) that refined statistical techniques are useful to distinguish the "out-of-control" situations.

Fig. 1 shows a graph of the diameters of thirty spindles produced successively on a lathe; the target value or specification diameter is 36 mm. The standard deviation of the diameters is 5 mm and the process

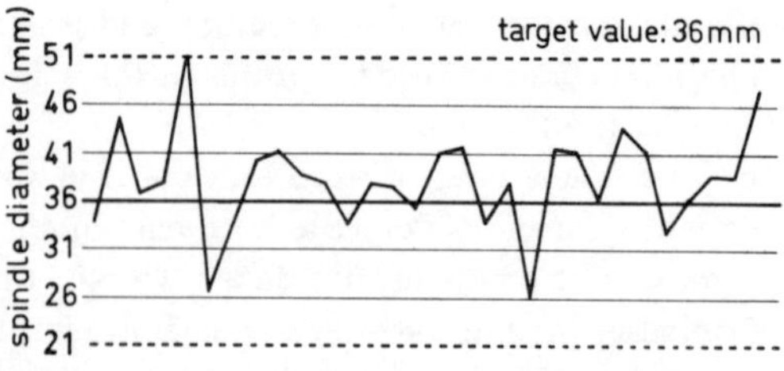

Fig. 1   Diameters of thirty spindles

is out of control: the average spindle diameter is 38 mm. As this represents a bias of only 40 per cent of the standard deviation of the diameters, the bias is hardly noticed in the graph.

The simplest type of control chart, the Shewhart chart, uses a graph like that of Fig. 1 on which the target line and two control lines have been drawn, one on either side of the target line. The distance between these control lines and the target line is taken as being proportional to the standard deviation ("sigma") of the random variation, the proportionality constant being chosen traditionally at three ("three-sigma limits"). In Fig. 1 the control lines are shown as dashed lines. These lines are often referred to as "action lines". Whenever an observation outside the region enclosed by the action lines occurs, the possibility of the process being out of control is assumed and appropriate action taken.

In the example of Fig. 1, the control chart does not trigger any alarm, because the peaks remain just below the upper action line. This illustrates the inability of the Shewhart chart to indicate slight inaccuracies promptly, although it is quite a satisfactory test if one is mainly interested in finding a large bias. In Table 1 we show, for a Normal distribution of the observations, the average number of obser-

Table 1   Average run lengths of Shewhart tests applied to normally distributed observations

A.R.L. for three-sigma limits	Bias (in sigma)	A.R.L. for two-sigma limits
385	0·0	22·0
200	0·4	15·9
71·4	0·8	8·50
27·8	1·2	4·70
12·4	1·6	2·90
6·30	2·0	2·00
3·65	2·4	1·52
2·38	2·8	1·27
1·73	3·2	1·13
1·38	3·6	1·058
1·19	4·0	1·023
1·023	5·0	1·0013

For the test with three-sigma limits the A.R.L. for a bias θ equals $1/(1-q)$,

in which $q = \dfrac{1}{\sqrt{2\pi}} \displaystyle\int_{-3-\theta}^{3-\theta} \exp(-x^2/2)\, dx.$

vations needed to trigger a query as a function of the bias. We speak here of the Average Run Length of the test as a function of the bias. It is seen that for a large bias the Average Run Length (A.R.L.) is small, i.e. these deviations will be found promptly. If one wants to improve the sensitivity of the control chart to a small bias, this can be effected by narrowing the region between the action limits. This will, however, result in more frequent false alarms, when the process is in control. These points are illustrated in Table 1, in which not only the A.R.L.'s for three-sigma limits, but also those for two-sigma limits are given for comparison.

We observe that in the case of the three-sigma limits there is one false alarm in every 385 observations (on average), whereas with the two-sigma limits there is one in every 22 observations. These are well-known facts. In each particular application one must take a decision on the proper compromise between too many false alarms and a too tardy reaction to real errors.

Because of the inability of the Shewhart test to detect a small bias quickly, several modifications of it have been introduced in the course of time. All these tests are based on the principle of the simultaneous use of several observations in a test. The simplest modification is the introduction of so-called warning limits in addition to the action limits – see Dudding and Jennett (1942). Traditionally the warning limits were taken at two sigma with the action limits at three sigma.

The rule for operating the control chart with warning limits is: whenever either a single observation lies outside the band enclosed by the action limits or two *successive* observations lie outside the band enclosed by the warning limits (at the same side of this band), an out-of-control signal is given. It has been remarked by Page that two successive observations outside the inner band, but on opposite sides of that band, could also be used to trigger a different kind of control signal, indicating that the standard deviation of the process exceeds the estimate currently used for it. We shall not consider this case further here.

This modified Shewhart test is triggered more quickly than the Shewhart test in the presence of small biases. This is illustrated in Table 2, in which the A.R.L.'s for the modified test are shown. The values of the limits have been set at 3·1 and 2·1 sigma for the action and warning limits, respectively, to give approximately the same A.R.L. at zero bias.

When comparing these values for the A.R.L. with those of Table 1 for the unmodified Shewhart test, we notice that the unmodified test reacts more quickly than the modified test only for a very large bias (larger than 4 sigma). This is because the action limit for the modified test had to be chosen slightly higher than that of the Shewhart test to give the same average number of false alarms.

An explanation for the success of the modified Shewhart test for

Table 2   Average run lengths for a Shewhart test with warning limits, for normally distributed observations (limits at 3·1 and 2·1 sigma)

Bias (sigma)	A.R.L.
0·0	390
0·4	187
0·8	57·2
1·2	20·16
1·6	8·78
2·0	4·65
2·4	2·90
2·8	2·06
3·2	1·61
3·6	1·35
4·0	1·195
5·0	1·029

For this test the A.R.L. for bias equals $(1+p)(1+r)\{1-pr-q(1+p)(1+r)\}^{-1}$,

in which $p = \int_{2\cdot1-\theta}^{3\cdot1-\theta} \exp(-x^2/2)\,dx, \quad q = \int_{-2\cdot1-\theta}^{2\cdot1-\theta} \exp(-x^2/2)\,dx, \quad r = \int_{-3\cdot1-\theta}^{-2\cdot1-\theta} \exp(-x^2/2)\,dx.$

small deviations is the fact that the combined information of the two most recent observations is being used. It should be remarked, however, that the Shewhart test, introduced in 1931, has dominated statistical practice for a long time and still plays an important role.

The cumulative sum test uses the combined information of any number of observations, in fact, of all the observations that have been obtained up to the time of testing. This is achieved by using the cumulative sum of deviations from target rather than the individual observations. In Fig. 2 a graph is given of the cumulative sums of differences

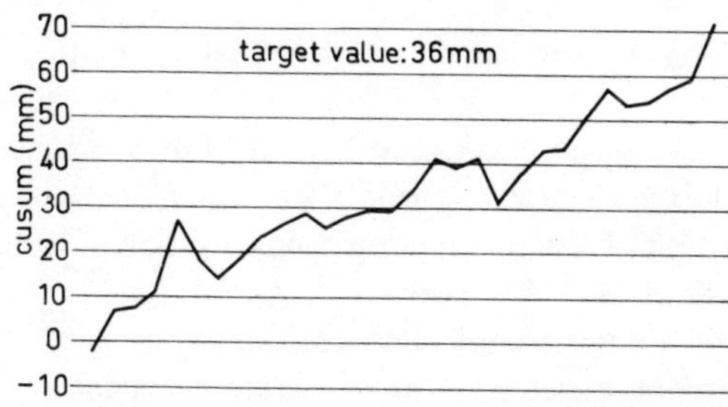

Fig. 2   Cumulative sums of deviations from target for spindle diameters

6

between the spindle diameters of Fig. 1 and their target value of 36 mm.
We see that even with the small bias of 0·4 sigma, there is a notice-
able tendency for the graph to rise rather than to fluctuate.

Obviously, the slopes between successive points on the cusum
graph are good indicators, if one wants to find significant biases. The
slope of a line drawn between any point on the graph and the last point
plotted is the average deviation from target for the observations between these points. We expect the control chart to give rise to a warning signal whenever this slope becomes too great, either in the positive or in the negative sense. Some part of the range of these slopes
may be ascribed to chance; but, naturally, as the number of observations involved increases, the range becomes narrower.

Page (1954a) and Ewan and Kemp (1960) have given procedures for
testing for significant increases in slope. These are discussed fully
in section 1.3. Barnard (1959) introduced an elegant graphical procedure which gives exactly the same results as the numerical procedure
mentioned above. We shall use the graphical method as a starting-point
in our discussion of cusum tests.

The graphical method uses a V-shaped mask, which is laid on top
of the cusum graph, so that the last point plotted in the graph is in a
marked position on the mask. This position is the centre of the trunca-

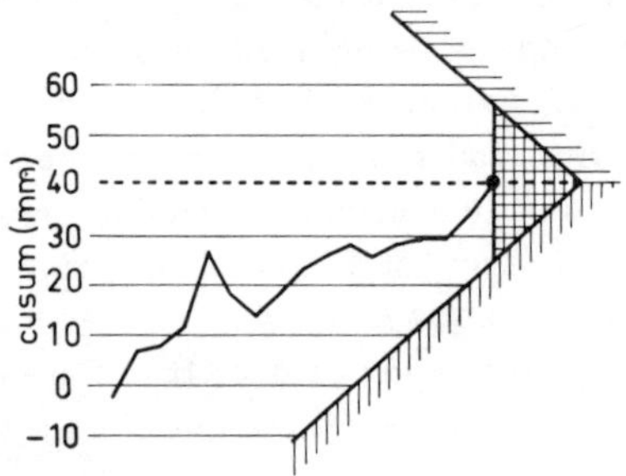

Fig. 3   Cusum graph with V-mask in position

ted end of the triangular mask; see Fig. 3. If at any time the cusum
graph cuts either edge of the V-mask, an out-of-control signal is given.
In Fig. 3 a section of the cusum graph of Fig. 2 is shown, with a V-mask
superimposed on it.

Clearly, for a short run of observations, it is possible to have a
greater average deviation from target, without cutting one of the edges
of the mask, than for a long run of observations. The limiting acceptable average deviation (for very long runs) is the slope of the legs of

the V-mask.

A control chart based on a cusum graph and V-mask is fairly easy to operate and has several advantages over conventional procedures:

(1) For the same number of false alarms, cusum tests can be made to react more promptly to a small bias.

(2) If an out-of-control situation is indicated, it can also be seen from the graph *when* the change first occurred.

(3) An estimate of the bias can be made by measuring the slope of the graph between the point at which the slope increases and the last point.

A disadvantage is that cusum graphs tend to consume a lot of graph paper by running off the top or bottom of the paper. Barnard (1959) has suggested the use of a cylindrical graph, which would be suitable because it is the slope rather than the level of the cusum that matters in the test.

In Table 3 the A.R.L. is given for the cusum test with the V-mask of Fig. 3, as a function of the bias. This particular cusum test can be

Table 3　Average run length for a cusum test (Normal distribution)

Bias (sigma)	A.R.L.
0·0	395
0·4	92·5
0·8	20·0
1·2	8·06
1·6	4·73
2·0	3·34
2·4	2·61
2·8	2·16
3·2	1·86
3·6	1·64
4·0	1·45
5·0	1·125

seen to react more quickly than either the Shewhart test or the modified Shewhart test, when the bias is less than 2 sigma (for those tests for which the A.R.L. is given in Tables 1 and 2). These values of the A.R.L. were obtained numerically by the author; see sections 2.2 and 3.7.

Looking at the V-mask more closely (Fig. 4), we see that the straight line V-mask has four parameters: two for the test on upward

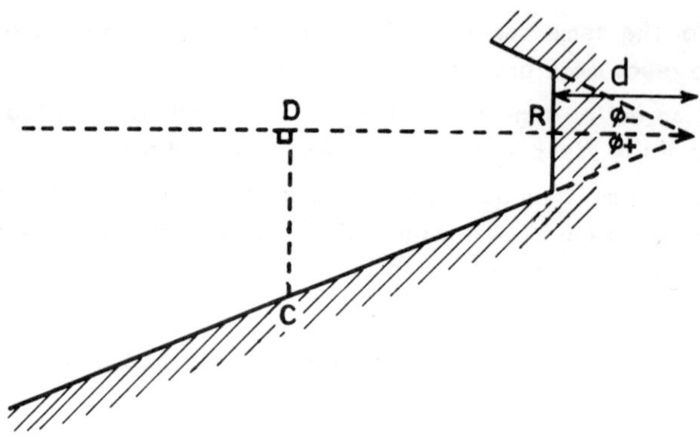

Fig. 4   Dimensions of the V-mask

changes and two for the test on downward changes. Often a symmetrical test is chosen, in which case only two parameters need be specified.

The parameters for a one-sided test are d, the distance from the truncated end of the V to the apex of the V, and ϕ, the angle between one leg of the mask and the horizontal. For symmetrical or one-sided tests we shall refer to d and ϕ; for asymmetrical tests to d^+, ϕ^+ and d^-, ϕ^-.

For given scaling of the graph, d and ϕ determine uniquely the properties of the cusum test. Practical information on scaling of the cusum graph is given in section 2.1.

As it is convenient to use one and the same mask for several applications, it pays to standardize the scales of the graph. The vertical axis can always be scaled in terms of the standard deviation of the random variation of the process. This will also improve the readability of the graph. The actual units of measurement can be added on an extra scale, if they are also wanted.

The boundaries of the mask shown above are straight lines. There is no need to restrict oneself to such simple forms. In fact, there are reasons to believe that segments of parabolas would be even better, but so far no analysis of the properties of such tests is available. We must also guard against exaggerated perfectionism in trying to improve a test procedure which is already very good. The argument in the following paragraphs indicates that no spectacular improvements are to be expected.

We shall first look more closely at the significance of the intersection of the graph with one of the edges of the mask at a particular point (see Fig. 4). To keep the formulas simple at this stage, we assume, up to section 2.1, that one standard deviation on the vertical scale and one time unit (distance between two observations) on the horizontal scale are both equal to the unit length.

Let the graph cut the mask at C. The number of observations between this point and the last point plotted is DR divided by the length between two observations on the horizontal scale. Therefore, the average upward slope g (expressed in standard deviations per observation), which is an estimate of the bias, is

$$g = \frac{CD}{DR}. \tag{1}$$

It follows from (1) that the mask will indicate a significant bias when

$$g > \frac{CD}{DR}. \tag{2}$$

Writing this in terms of the V-mask parameters one learns that a signal will occur if

$$g > \frac{(DR + d)\tan \phi}{DR}. \tag{3}$$

(3) shows that the cusum test is effectively an aggregate of several tests (for $DR = 1, 2, ..., N$) applied simultaneously. Writing T for DR, we see that a signal will occur whenever at least one of the slopes over a span of T observations exceeds $\{(T + d)\tan \phi\}/T$, for $T = 1, 2, ..., N$.

In Fig. 5 the maximum permitted slopes for spans of up to 90 observations are given for a particular cusum test, parameter values being $d = 11\cdot5$ and $\tan \phi = 0\cdot2$. From (3) we see that, amongst other things, this amounts to allowing a single deviation of up to $2\cdot5\sigma$ without a signal. Hence the Shewhart test with limit $2\cdot5\sigma$ is one of the tests implicitly applied, and therefore the A.R.L. cannot be greater than that of that Shewhart test.

For an increasing number of observations the maximum tolerated mean deviation of the sample decreases; however, the standard deviation of the sample mean also decreases, being proportional to $1/\sqrt{T}$.

This standard deviation is shown in Fig. 5 for comparison. Apparently the hyperbola of the maximum permitted slopes can be approximated fairly well in the region shown by the inverse square root of the span.

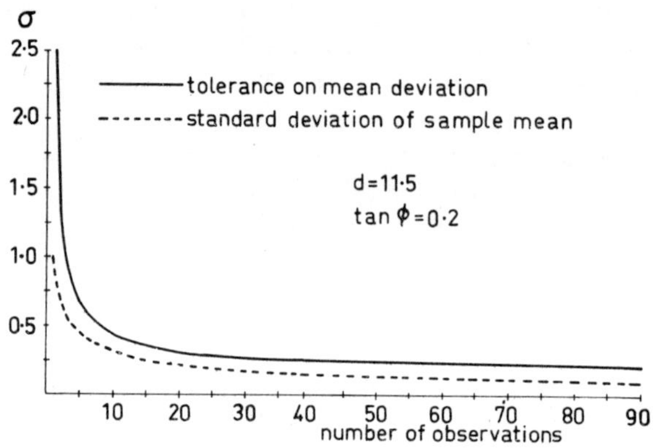

Fig. 5   Comparison of tolerated mean deviation and its standard deviation

For the example of Fig. 5, we have calculated the ratio of 2·5 times the standard deviation of the sample mean and the right-hand side of expression (3). For $T = 1$ this is equal to 1 by definition. It is found that the ratio reaches a maximum value 1·84 for $T = 11·5$ and is equal to 1 again for $T = (11·5)^2 = 132·25$, after which the ratio approaches zero as an inverse square root. We have used a continuous variable for T here for simplicity's sake; the correct values for T are one of the two nearest integers to 11·5 and 132·25, respectively.

In general, the maximum of the ratio is always reached for $T = d$, and its value is $(1 + d) \sqrt{d}/2d$. The value 1 is assumed for $T = 1$ and $T = d^2$. Accordingly, the lower d is, the more closely the hyperbola resembles the inverse square root, but also the shorter is the span of good approximation.

This relationship between the tolerance on mean deviation in a sample and the standard deviation of that sample is a heuristic argument in favour of the cusum test; however, as is the trouble with heuristic arguments, this does not prove anything. It does indicate, though, that little improvement is to be expected from changing the straight-line mask into a mask with parabolic limits.

1.2 Applications of cumulative sum tests

In principle, cumulative sum tests may be applied in all cases where the Shewhart test is traditionally applied, such as in quality control. In addition to this there may be useful applications when the number of available observations is very small or the interval between observations long, which makes efficiency of the test of the utmost importance. An example of this is the control of forecast-errors of a sales forecasting procedure; the observations are then naturally spread out in time and cannot be obtained more frequently without incurring great expense. It is in those cases especially that the improved effectiveness in signalling real changes shows to full advantage, and the extra work involved in a more complicated procedure is justified.

Often in statistical quality control improved effectiveness can be attained just as easily by increasing the sample size and keeping the simple Shewhart test. Also one can usually increase the frequency of sampling. If, in process control, one wants to obtain an estimate of the approximate magnitude of the change and the time at which the change occurred along with an out-of-control signal, the graphical cusum test takes precedence; as soon as a signal has occurred, a straight line can be drawn through the last portion of the graph and a second line through the part of the graph made before the change occurred. The point where these two lines meet is the estimated time of occurrence of the change. Also the slope of the line through the later portion of the graph provides an estimate of the magnitude of the change.

In the author's opinion, cusum tests have a great future ahead in the field of routine sales forecasting, particularly in large systems for many products. In these systems, which are often computerized, there is little human checking and there is a natural data interval which cannot easily be shortened. The cusum test can fulfil the role of a filter to single out those series which merit attention. The term "management by exception" is sometimes used rather grandly in this context.

A special case is forecasting by cusum test. This is a step further in computerized forecasting systems, in which human intervention is almost completely eliminated. An initial forecast is entered for each item. This forecast may consist generally of a number of parameters of a trend line, or may just be a single constant. The initial forecast is then maintained until a cusum test indicates an out-of-control situation (usually there are two tests for each item: one to test for positive deviations and one for negative deviations). When a cusum exceeds its limits, some standard rule for re-estimating the parameters is activated, and the cusum tests are reset to zero. The forecasts do not change in the majority of review cycles. This often saves a lot of computation in production plans, etc. which are based on the forecasts. (One might have to solve an integer linear program each time a forecast is changed.) Naturally one needs a numerical version of the cusum test; this is described in the next chapter.

An alternative method to estimate the new mean is the following. A parabolic mask shaped roughly like the V-mask is constructed and, like the V-mask, its apex is made to coincide with the last point plotted on the cusum graph. The mask is then rotated around this point until the number of successive points inside the confines of the mask is maximized. The slope of the centre line of the mask is then an estimate of the average deviation from target.

Truax (1961) observes that cusum tests in their graphical form have an even greater advantage over Shewhart charts than one would deduce from the respective A.R.L.'s. The amount of visual estimation power is such that with properly proportioned cusum charts it is possible for the people operating the process to develop a "feel" for its behaviour — which cannot easily be acquired when using standard Shewhart charts.

1.3   Graphical and numerical tests

The cusum graph has great virtue, deriving from its easy visual interpretation. However, for computer applications, a numerical version is required. Other applications can also be thought of where it would be too cumbersome to construct and inspect graphs for every item. From (3) in section 1.1 a possible numerical procedure follows: one can compute the partial sums of deviations from target over all possible time spans and use relation (3) to test each sum for significance. This would involve rather a lot of computation and also of storage space in computer applications, of which the latter is often the more important consideration.

Fortunately, for the straight-line boundaries of the V-mask as assumed in (3), there is a simpler numerical procedure, which is equivalent to the graphical procedure.

From each standardized deviation from target a constant is subtracted, which is called the *reference value*. The resulting figures are then accumulated as an ordinary cusum. The effect of subtracting the reference value is to produce a downward slope of the cusums, when the process is in control. If, however, the bias exceeds the reference value, a positive tendency persists in the now modified cusum graph.

As soon as this biased cusum graph exceeds its minimum by more than a specified amount, for which we shall use the symbol h, an out-of-control signal is triggered. h is called the *decision interval*. The symbol used for the reference value is k. In this monograph, both h and k will always be expressed in standard deviations of the observations. We shall show later that the above procedure is equivalent to that in which a straight-limb V-mask is applied to the ordinary cusum graph; the relationship between d and ϕ, the V-mask parameters, and the parameters k and h, will follow from the proof.

The above procedure is still in graphical terms. It is, however, easily applied both graphically and numerically. Because a signal only occurs when the previous minimum is exceeded by more than h, it is not necessary to plot the graph of biased cusums as long as the graph is sloping downwards. One has only to start plotting as soon as a positive contribution to this sum occurs, i.e., if the observed deviation minus the reference value is positive. One should then continue plotting until the sum either exceeds the decision interval h, or reaches zero again. After reaching zero, no further plotting of the sum is necessary until the next positive contribution.

Let S_t be the value of the test quantity plotted after the tth observation and e_t the difference between the tth observation and the target value. Then the test procedure may be written as

$$\left.\begin{aligned} S_t &= \max(0, S_{t-1} + e_t - k) \\[2mm] S_0 &= 0 \end{aligned}\right\} \tag{4}$$

$$\text{signal if } S_t > h. \tag{5}$$

This test procedure will only detect positive biases. To detect negative biases as well, which in the graphical procedure is done by the second limb of the V-mask, it is necessary to have a second, similar test in which a reference value is added to each standardized deviation and a signal triggered off as soon as the sum falls below the previous maximum by more than the decision interval.

To avoid confusion about the parameters of these two separate tests, the symbols h^+, k^+ and h^-, k^- will be used for tests on positive and negative biases, respectively. The operating formulas then become

$$S_t^+ = \max(0, S_{t-1}^+ + e_t - k^+) \tag{6}$$

$$S_t^- = \min(0, S_{t-1}^- + e_t + k^-) \tag{7}$$

and a signal is triggered either because S^+ exceeds h^+ or because S^- falls below $-h^-$. Whenever no confusion can arise, we shall still be using the symbols S, h and k. In the case of a symmetrical V-mask, the parameters of the negative test must of course be the same as those for the positive test.

Proof of equivalence of graphical and numerical procedure

From (3) we find that the out-of-control condition is characterized by at least one of the sums $\sum_{i=0}^{T-1} e_{t-i}$ (for $T = 1, 2, \ldots, t$) exceeding $(T + d)\tan\phi$.

Or the out-of-control condition at time t means that one of the inequalities (8) is satisfied:

$$\sum_{i=0}^{T-1} (e_{t-i} - \tan\phi) > d\tan\phi \quad (T = 1, 2, \ldots, t). \tag{8}$$

Substituting $k = \tan\phi$ and $h = d\tan\phi$ into (8), this becomes

$$\sum_{i=0}^{T-1} (e_{t-i} - k) > h, \text{ for some } T (1 \leqslant T \leqslant t). \tag{9}$$

The process is thus considered to be in control if for no T the inequality (9) is true; this is the same as saying that the current point in the graph of biased cusums should not be more than h above the lowest point of this graph. This proves the equivalence of the two modes of cusum test with the parameter correspondence

$$h = d\tan\phi \tag{10a}$$

$$k = \tan\phi = h/d. \tag{10b}$$

The numerical version characterized by (4) and (5) differs from this only in that the latest value of the minimum cusum is always equated to zero. The advantage of this is that it is convenient graphically and numerically, because the values taken on by the sum are bounded on a finite interval and only the last value of the sum has to be remembered.

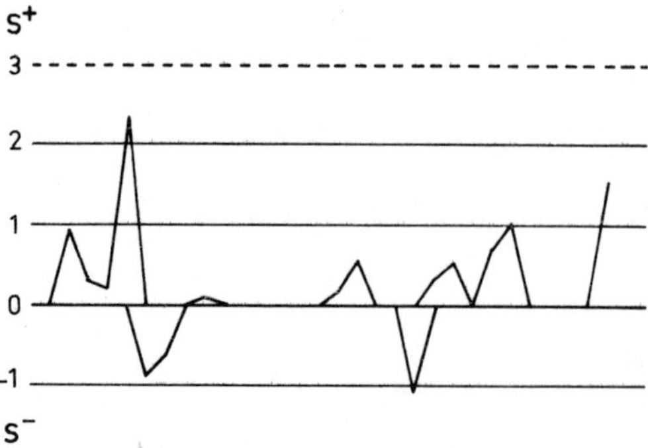

Fig. 6   Numerical cusum test quantities S^+ and S^-

Fig. 6 shows this type of cusum graph for the same data as were used for Fig. 2. The cusum parameters used are $k = 0{\cdot}85$ and $h = 3$. These parameters correspond to the cusum test for which the A.R.L. has been given in Table 3. The cusum tests for positive and negative biases are both shown.

1.4   The cusum test as a sequential test applied in reverse

As the popularity of the cusum test gained momentum, the need for accurate information about its properties arose. Most authors reported that they determined the best shape of the V-mask to use empirically, adjusting the mask by "cutting and trying" until there were neither too many false alarms when the process was in control, nor very long runs of observations during bad quality production.

Johnson (1961) suggested using the precise analogy of cusum tests and standard sequential probability ratio tests to estimate the average run length. The quality of these estimates ranges from fair to misleading. The accuracy of several approximate formulas is discussed in Part Three of this monograph.

The theory of sequential tests can be found in Wald (1947) and in books on industrial experimentation such as Davies (1963). We here summarize the procedure and terms used.

Suppose we want to discriminate between two hypotheses about a variable x, the null hypothesis being that the process is in control, and the alternative hypothesis being that the process variable is biased by g standard deviations. Let us give the symbols H_0 and H_1 respectively to these hypotheses.

Then calculate at each stage of sampling the ratio of the likelihoods of obtaining the observed values under H_1 and H_0 respectively. If this ratio R is large, the alternative hypothesis H_1 is accepted; if the ratio is small, the null hypothesis H_0 is accepted; and thirdly, if R has an intermediate value, the decision is delayed until further observations have produced either a higher or a lower R.

It is desirable to choose the limits for the ratio R such that the probabilities of making the two possible wrong decisions are small. Let us say that we want the probability of creating a false alarm (i.e. accept H_1 when H_0 is true) to be equal to α_0, and the probability of accepting the null hypothesis, when the process is really out of control by the amount g, to be α_1.

These probabilities will approximately obtain if the rules used are —

$$\text{if} \quad R \leqslant \frac{\alpha_1}{1-\alpha_0} \qquad\qquad \text{accept } H_0$$

$$\text{if} \quad R \geqslant \frac{1-\alpha_1}{\alpha_0} \qquad\qquad \text{accept } H_1 \qquad\qquad\qquad (11)$$

$$\text{if} \quad \frac{\alpha_1}{1-\alpha_0} < R < \frac{1-\alpha_1}{\alpha_0} \qquad \text{continue sampling}$$

If the observations have a Normal distribution, and if H_0 specifies a mean μ, and H_1 specifies a mean $\mu + g\sigma$, the probability likelihood ratio of a sample of N observations $x_1, x_2, \ldots, x_N$ is

$$R = \frac{(\sigma\sqrt{2\pi})^{-N} \exp\left\{-\dfrac{1}{2\sigma^2} \sum_{i=1}^{N} (x_i - \mu - g\sigma)^2\right\}}{(\sigma\sqrt{2\pi})^{-N} \exp\left\{-\dfrac{1}{2\sigma^2} \sum_{i=1}^{N} (x_i - \mu)^2\right\}}$$

$$= \exp\left\{\frac{g}{\sigma} \sum_{i=1}^{N} (x_i - \mu) - \frac{1}{2}Ng^2\right\} . \qquad\qquad (12)$$

Applying relations (11) to this ratio and simplifying, the test procedure becomes

$$\text{if} \quad \frac{1}{\sigma} \sum_{i=1}^{N} (x_i - \mu) \leqslant \frac{1}{g} \ln \frac{\alpha_1}{1-\alpha_0} + \frac{1}{2}Ng \quad \text{accept mean } \mu$$

$$\qquad\qquad\qquad\qquad\qquad\qquad\qquad\qquad\qquad\qquad\qquad (13)$$

$$\text{if} \quad \frac{1}{\sigma} \sum_{i=1}^{N} (x_i - \mu) \geqslant \frac{1}{g} \ln \frac{1-\alpha_1}{\alpha_0} + \frac{1}{2}Ng \quad \text{accept mean } \mu + g\sigma$$

and continue taking observations until one of these two decisions has been reached.

The left-hand sides of (13) are the normalized cumulative sums of deviations from the null hypothesis average or target mean. If we plot these against the number of observations N, the inequalities (13) can be drawn as two straight-line boundaries; if the cusum is between these boundaries, further sampling is deemed necessary. The region between the two boundaries is therefore called the "continuation region".

The connection between the cusum test and the sequential probability ratio test is now clear, especially if we consider the two-sided test procedure proposed by Armitage (1950), who also considers the third hypothesis H_{-1} that the true mean is $\mu - g\sigma$. A graph of the two combined sequential tests obtained for that case is shown in Fig. 7.

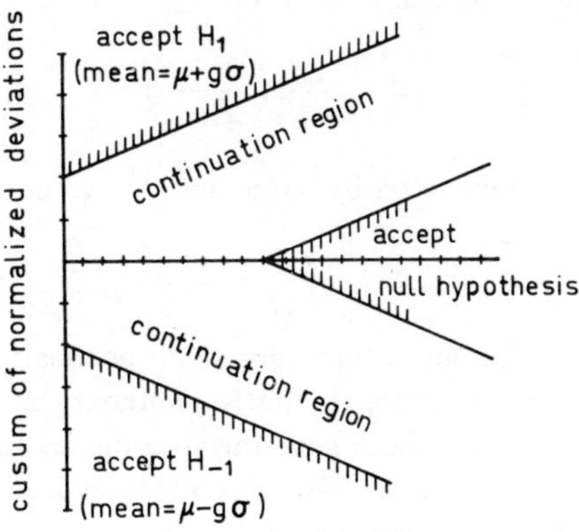

Fig. 7   Two-sided sequential acceptance test

A cusum test with a V-mask is thus seen to be a double sequential test applied backwards, i.e. the last observation is always used (= the difference between the last two points of the cusum graph); the observation before that is also used if the last observation did not lead to a decision about the validity of the null hypothesis, and so on until either a decision is reached or one runs out of observations. A slight difference in interpretation between the double sequential test and the cusum test is that as the number of observations is not controlled by the test procedure, the region in which H_0 is accepted melts into the continuation region. The null hypothesis is either rejected at one of the two boundaries or it is considered to describe the situation better than either of the alternative hypotheses. The null hypothesis is then "not rejected" at the point in question, but sampling continues and a further sample may still prove it wrong.

The sequential test will always end by accepting either H_0 or H_1 or H_{-1} (see Wald (1947)). The cusum test will always sooner or later hit one of the boundaries and will thus always ultimately lead to rejection of the null hypothesis.

It is possible to exploit this analogy to determine the parameters d and ϕ of the V-mask of the cusum test to be used. First we specify a bias $g\phi$, which should be detectable with fair certainty (this is sometimes called "rejectable quality level", RQL). Using (13), the slope of the legs of the V-mask should then be

$$\phi = \tan^{-1}\frac{1}{2}g. \tag{14}$$

The lead distance of the mask becomes

$$d = \frac{2}{g^2} \ln \frac{1 - \alpha_1}{\alpha_0} . \tag{15}$$

This can be approximated by assuming α_1 to be small, as

$$d = -\frac{2}{g^2} \ln \alpha_0. \tag{16}$$

The last approximation is not only made because of algebraic convenience. As indicated above, the null hypothesis is never accepted, so the parameter α_1 is without meaning in a cusum test. The parameter α_0 represents the acceptable proportion of samples, in a single-sided test, which trigger false alarms.

The theoretical framework described above gives a general way of selecting cusum test parameters. We can use the calculated values as starting-points for the "cut and try" methods to find the parameters that will actually be used.

The use of equations (14) and (16) rests on the following two assumptions:

(1) Although α_0 will not be exactly the proportion of false alarms, tests with the same α_0 have approximately the same proportion of false alarms; that is, these tests have approximately the same average run length when the null hypothesis is true.

(2) The average run length, if the bias is $g\sigma$, is $-2/g^2 \ln \alpha_0$, which is equal to $-\ln \alpha_0 / \mathcal{E}[R|H_1]$, see Johnson (1961). This approximation is also equal to $d = h/k$. For further comments on this type of approximation see section 3.5.

The numerical cusum parameters are

$$\left. \begin{aligned} h &= d \tan \phi = -\frac{1}{g} \ln \alpha_0 \\ k &= \tan \phi = \tfrac{1}{2} g \end{aligned} \right\} \tag{17}$$

Summarizing, we can say that the analogy with sequential analysis can be useful when first choosing the parameters to be used. One has to be careful in drawing conclusions from calculations thus made, because one has no information about the accuracy of assumptions (1) and (2).

The most striking general result of this section seems to be that detection of a shift of g standard deviations is best done by using $k = \tfrac{1}{2}g$, independently of what the other parameters are. This is referred to in the literature as choosing a "central reference value".

1.5 Average sample number; operating characteristic; average run length

In this section we discuss those quantities which define the behaviour of a cusum test in such a way that comparison of various tests is possible.

The terminology stems in part from sequential analysis. The term "average run length" (A.R.L.) is peculiar to process control charts.

In a sequential acceptance procedure a test always ends with acceptance or rejection of the batch tested. The number of observations before a decision is reached is variable and depends on the actual outcomes of the samples. The number of observations in the whole sample up to the decision is the sample number. The sample number is a random variable. The mean of the distribution of this random variable is termed the *average sample number*.

The probability that the test will end by accepting the null hypothesis is termed the *operating characteristic* of the test.

For a cusum test the values of the average sample number and operating characteristic depend on the shape and parameters of the distribution of the observations, and also on the value of the numerical cusum (formula (4)) at which the test is started.

Usually a test will be started with the cusum set at zero, but it is possible that one would like to know the average sample number and operating characteristic when halfway into a test; the information obtained up to the point reached is then contained in the numerical value of the cusum. The expected sample number for the whole test is the number of observations taken so far, plus the conditional expectation of the number of observations to come for a given value of the cusum now.

Let the value of the cusum be z. We then use the symbol

$$N(z; p_1, p_2, \ldots, p_n)$$

to denote the average sample number; the p_i are the parameters of the distribution of the control variable and of the cusum test. Similarly, we use $P(z; p_1, p_2, \ldots, p_n)$ for the operating characteristic of a test starting at z.

The average sample number and operating characteristic are sufficient to describe the properties of an acceptance scheme in terms of expectations. In the application of a cusum test to process control, however, the null hypothesis is never accepted, as we have remarked before. If the operating characteristic is less than 1, which is the case with any probability distribution of the observations which has not degenerated, a cusum test will eventually lead to rejection of the null hypothesis. The number of observations up to this rejection of H_0 is called the *run length*. This is also a random variable with the *run length distribution* as its distribution. The mean of the run length distribution is the *average run length* of the test. We use the symbol $L(z; p_1, p_2, ..., p_n)$.

For the cusum test the average sample number is the expected number of observations until the cusum either returns to zero or exceeds the decision interval h. The operating characteristic is the probability that a return to zero occurs, rather than a value above h.

A run of observations is a succession of tests of which all but the last return to the zero line. The average run length must therefore be the average sample number multiplied by the expected number of tests before one exceeds the decision interval. The probability that one particular test exceeds h is $1 - P(0; p_1, p_2, ..., p_n)$. The average number of tests is then $1/(1 - P)$ and the distribution of the number of these tests is a geometric one, the probability that m tests will be made before a departure from target is noticed being

$$\text{Prob (number of tests} = m) = (1 - P)P^{m-1}. \qquad (18)$$

It is now clear that the average run length for a test starting at zero is connected with the average sample number and operating characteristic by

$$L(0; p_1, p_2, ..., p_n) = \frac{N(0; p_1, p_2, ..., p_n)}{1 - P(0; p_1, p_2, ..., p_n)}. \qquad (19)$$

For starting values z other than zero the average run length is the average sample number $N(z)$ plus the average run length from $z = 0$ weighted by the probability $P(z)$ that this test will return to zero. Hence

$$L(z) = N(z) + L(0)P(z). \qquad (20)$$

In the next section we show that the three quantities defined here satisfy integral equations which are closely related to one another. The treatment of run length distribution and the solution of the integral equations is left to Part Three of the monograph.

The average run length (A.R.L.) of a two-sided test is a function of the A.R.L's of the positive and negative test. Let us call these L^+ and L^- respectively. L denotes the A.R.L. of the two-sided test. Consider a very long run of observations of length m. The expected number of signals given by the positive test is then m/L^+; similarly, the expected number of signals from the negative test is m/L^-. These numbers of signals are additive if the two tests are exclusive, i.e. if one of the tests signals, the other test should not be in a state from which a signal could have resulted at a later stage. In the case of the cusum test this means that the numerical cusum of the negative (positive) test should be zero, whenever the positive (negative) test signals. In the case of the two-sided Shewhart test with warning lines, it implies that the warning line of the positive test should lie above the warning line of the negative test. With the ordinary Shewhart chart this condition is always satisfied.

If, then, the condition is satisfied and hence the number of signals is additive, the total number of signals m/L is equal to $m/L^+ + m/L^-$, which gives

$$L = \frac{L^+ L^-}{L^+ + L^-} .\tag{21}$$

If the condition is not satisfied, we have

$$L \geqslant \frac{L^+ L^-}{L^+ + L^-} .$$

In the Appendix it is proved that the condition is always satisfied for a symmetrical cusum test. Also, it is shown that the condition is satisfied for asymmetrical cusum tests, if and only if

$$k^+ + k^- \geqslant |h^+ - h^-|\tag{22}$$

or in terms of the mask parameters

$$\tan \phi^+ + \tan \phi^- \geqslant |d^+ \tan \phi^+ - d^- \tan \phi^-|\tag{23}$$

which becomes, if $d^+ = d^- = d$,

$$\sin(\phi^+ + \phi^-) \geqslant d|\sin(\phi^+ - \phi^-)|.\tag{24}$$

As previously mentioned, the A.R.L. depends on $p_1, \ldots, p_n, n-2$ of which are distribution parameters. Usually one attempts to test only one of these parameters at a time against a target value. We shall use the symbol θ for the departure from target (bias) of the parameter tested, and express this in standard deviations of the sample values of this parameter.

The graph of the A.R.L. as a function of θ, the $L(0; \theta)$ plot, is usually produced with a logarithmic scale for the A.R.L. Fig. 8 shows this for the data of Tables 1, 2 and 3.

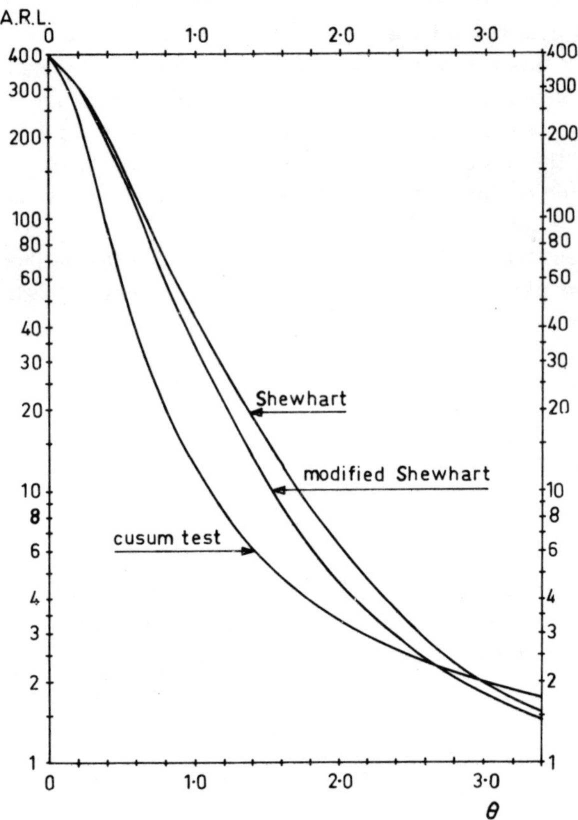

Fig. 8   A.R.L. for the three tests of Tables 1, 2 and 3

1.6   The integral equations

Up to this point we have defined average sample number, operating characteristic, and average run length without mentioning how the
numerical values for the tables were obtained; this is the subject of
Part Three of the monograph. In Part Two the properties of cusum
tests are treated in some detail, without attempting to justify the numerical values in the tables. The main tables for reference are given
in Part Two. If one is only interested in the application of cusum
tests, a reading of Part One and relevant sections of Part Two is sufficient, and Part Three can be skipped completely.

In the present section the integral equations that are satisfied by
the A.R.L., the average sample number, and the operating character-

istic are derived.

Let the distribution function of the values y to be accumulated in the numerical cusum test (= standardized deviations from target minus k) be $F(y; \mu_0, \theta, k)$. μ_0 is the target mean and θ the mean deviation from target. In the case of the normal distribution there is only one parameter, namely $\theta - k$, as

$$y_i = x_i - \mu_0 - k = x_i - \mu + \theta - k.$$

Then the expected run length of a test, starting at z (the cusum) satisfies the integral equation

$$L(z) = 1 + L(0) \int_{-\infty}^{-z} dF(\xi) + \int_{-z}^{h-z} L(z+\xi)\, dF(\xi) \quad (0 \leqslant z \leqslant h). \quad (25)$$

In these and subsequent equations we have left out the parameters about whose presence no confusion can arise, so for $L(z)$ read $L(z; \mu_0, \theta, k, h)$ similarly for $P(z)$ and $N(z)$.

The integrals in (25) are to be considered as Stieltjes integrals, thus admitting both discrete and continuous random variables. For a continuous variable we can then write, using $f(y)$ for the derivative of $F(y)$,

$$L(z) = 1 + L(0) \int_{-\infty}^{-z} f(\xi)\, d\xi + \int_{-z}^{h-z} f(\xi)\, L(z+\xi)\, d\xi \quad (0 \leqslant z \leqslant h). \quad (26)$$

This integral equation admits the following verbal, and therefore necessarily not quite exact, description: the expected run length of a test, which is now at z, equals 1 (the next observation, which is always there for $0 \leqslant z \leqslant h$) *plus* the probability that the next observation will return the cusum to zero multiplied by the expected run length from $z = 0$ *plus* the integral over the probabilities that the cusum lands somewhere between zero and h multiplied by the respective expected run lengths from the new value of the cusum ($= z + \xi$). (26) can be improved by change of variable ξ to $x - z$:

$$L(z) = 1 + L(0) \int_{-\infty}^{0} f(x-z)\, dx + \int_{0}^{h} f(x-z)\, L(x)\, dx$$

$$= 1 + L(0)\, F(-z) + \int_{0}^{h} f(x-z)\, L(x)\, dx \quad (0 \leqslant z \leqslant h). \quad (27)$$

We are usually only interested in the A.R.L. of a test starting at zero; knowledge of $L(0)$ would be sufficient. But, as is evident from (27), $L(z)$ must be known for all values of z in the interval $(0, h)$ to enable one to find $L(0)$. As (27) is not of a form which can be solved analytically in general, it will often be necessary to use numerical or approximate results. These are discussed in Part Three.

A lower bound on $L(0)$ may be readily found from (27) by substituting $z = 0$:

$$L(0) = 1 + L(0) F(0) + G(0),$$

where $G(0)$ is an unknown non-negative quantity. Thus

$$L(0) = \frac{1 + G(0)}{1 - F(0)} \geqslant \frac{1}{1 - F(0)}. \tag{28}$$

Similarly, an upper bound may be found for $L(0)$. As $L(z) \leqslant L(0)$ for $0 \leqslant z \leqslant h$,

$$G(0) \leqslant L(0) \int_0^h f(x - z)\, dx$$

hence

$$L(0) \leqslant 1 + L(0) F(h)$$

or

$$L(0) \leqslant \frac{1}{1 - F(h)}. \tag{29}$$

An interpretation of (29) is that the cusum test with parameters h and k has an A.R.L. not greater than that of a Shewhart test with limit $h + k$, which is obvious because such a Shewhart test is automatically included in the cusum test.

Unfortunately, these inequalities (28) and (29) do not usually set very tight limits on the value of $L(0)$. For a Normal distribution with $\mu = 0$ and cusum test parameters $k = 0 \cdot 2$ and $h = 2 \cdot 3$, we have from tables of the Normal probability integral

$$F(0) = 0 \cdot 5793 \qquad F(h) = 0 \cdot 9938.$$

Substituting these values into (28) and (29) gives the following limits on the A.R.L. of the cusum test:

$$2 \cdot 38 \leqslant L(0) \leqslant 161 \cdot 3.$$

Obviously such wide limits are not going to be very helpful. In this particular example the tightness improves a little for higher μ.

Better inequalities may be found for the Normal distribution in Kemp (1967b); they can also be derived for more general distributions from the formulas in Kemp (1967a).

The average sample number and operating characteristic satisfy integral equations, which are quite similar to (27). These are

$$N(z) = 1 + \int_0^h f(x - z) N(x)\, dx \qquad (0 \leqslant z \leqslant h) \tag{30}$$

$$P(z) = F(-z) + \int_0^h f(x - z) P(x)\, dx \qquad (0 \leqslant z \leqslant h). \tag{31}$$

As we have from the previous section the relation (19) between L, P and N, we need only solve two of the three integral equations. In Part Three it is shown that it is only necessary to solve one related integral equation (the resolvent equation) in order to solve all integral equations with the general appearance of (27), (30) and (31).

For discrete variables the integral equations become summation equations, but no essentially new points arise.

To illustrate the integral equations and their solutions, but without showing how the solutions are obtained, we conclude this section with two simple examples in which an analytic solution to the integral equations is known. The probability distributions are a uniform and an exponential distribution, respectively.

Example 1

Distribution: uniform between 0 and 2.

Cusum test parameters: $k = 1$ and $h = 1$.

Hence $f(x) = \frac{1}{2}$ $(-1 \leqslant x \leqslant 1)$ and $f(x) = 0$ everywhere else.

(30): $\quad N(z) = 1 + \int_0^1 \frac{1}{2} N(x)\, dx = 2 \quad (0 \leqslant z \leqslant 1)$

(31): $\quad P(z) = \frac{1}{2}(1 - z) + \int_0^1 \frac{1}{2} P(x)\, dx = 0{\cdot}75 - 0{\cdot}5z$

(19): $\quad L(0) = 8$

(20): $\quad L(z) = 8\,P(z) + N(z) = 8 - 4z$.

Example 2

Distribution: exponential with mean equal to unity.

Cusum test parameters $k = 1$ and $h = 1$.

For $\mu_0 = 1$, $\theta = 0$: $f(x) = \exp(-x - 1)$, $(-1 \leqslant x)$.

The solutions of (30) and (31) are:

$$N(z) = 1 + e^{-1} \int_0^1 \exp(z - x)\, N(x)\, dx = 1 + e^{-1} \exp(z)$$

$$= 1 + 0{\cdot}368 \exp(z)$$

$$P(z) = 1 - e^{-1} \exp(z) + e^{-1} \int_0^1 \exp(z - x)\, P(x)\, dx$$

$$= 1 - \frac{e^{-2}}{1 - e^{-1}} \exp(z)$$

$$= 1 - 0{\cdot}214 \exp(z)$$

$$L(0) = 6{\cdot}39$$

$$L(z) = 7{\cdot}39 - \exp(z).$$

2 APPLICATIONS

2.1 Practical hints for the application of cumulative tests; scaling

Once we have decided to use cumulative sum tests, there are a number of practical difficulties to be overcome. The first question is: What should the cusum parameters be? A second question, arising if the graphical mode is employed, concerns the scaling of the graph and V-mask.

If only the graphs of S^+ and S^- are plotted, scaling is relatively unimportant, but in the case of the cusum graph with the V-mask the possibility exists that the graph may run off the top or bottom of the page.

To come first to the choice of parameters, the user will have to specify what sort of performance he expects of the test. Often the user can specify a point at which he considers the process to be out of control in an unacceptable way. If the process is running at that level, a signal is required as quickly as possible, i.e. a low value of the A.R.L. at that point is desirable.

Let the standard deviation of the observed variable be σ. Let the level of unacceptable bias be $\delta\sigma$. In the terminology of section 1.5, the object is to obtain as small as possible an $L(0;\delta)$ for a given $L(0;0)$. $L(0;0)$ is defined as the inverse of the relative frequency of false alarms that is acceptable to the user. Another approach is to fix the A.R.L. at the unacceptable level to $L(0;\delta)$ and try to maximize $L(0;0)$ by the choice of parameters. The second method may be regarded as a dual to the first method.

We assume that the problem has been specified in one of the two ways indicated above. If $L(0;\theta)$ is known for a sufficiently extensive range of θ and the parameters h and k, there is no serious problem (supposing that there is a unique extremum for $L(0;\delta)$ or $L(0;0)$).

For some cases tables exist or can easily be produced; one of these cases is that in which the mean of a Normal distribution is controlled. We give fairly extensive tables for this important case in the next section. From these tables one can find the "optimum" parameters by a process of repeated inverse interpolation. It is usually found that the minimum $L(0;\delta)$ or maximum $L(0;0)$ does not depend too critically on the particular combination of h and k. This combination must, as a matter of course, satisfy the condition of giving the desired $L(0;0)$ or $L(0;\delta)$, respectively.

27

In Table 4 an example is shown for $\delta = 1$ and a desired average run length at the rejectable level of $5 : L(0;1) = 5$. The example refers to

Table 4   "Best test" for $L(0;1) = 5$; Normal distribution

h	$\theta - k$	k for test	$L(0;0)$
1·5	0·219	0·781	46·7
2·0	0·409	0·591	52·6
2·5	0·564	0·436	*** 52·9 ***
3·0	0·699	0·301	48·3
4·0	0·942	0·058	27·0
5·0	1·170	−0·170	21·1
6·0	1·393	−0·393	15·0

the Normal distribution. The value of $\theta - k$, for which the condition $L(0;1)$ is satisfied for the h of column 1, is given in column 2 of the table. The third column gives the value of k to be used to satisfy the restrictive condition. The last column then shows the average run length at $\theta = 0$. The maximum is reached for $h = 2{\cdot}5$ and $k = 0{\cdot}436$, i.e. a practically "central reference value", as the theory of sequential analysis recommends (section 1.4). It is clear from the figures that the maximum is very flat, so that it is not necessary to keep too closely to the values found.

From this example, based on accurate A.R.L. figures, we conclude that it seems quite safe to use central reference values, if the problem is posed. Hence k is equated immediately to $\frac{1}{2}\delta$. From the tables one then has to find the value for h which will make the side condition true for the given choice of k.

If no tables or approximations are available, or if there is no time for analysis to give parameter values of the "best test", we recommend the use of the approximations provided by sequential analysis as shown in section 1.4. Two quantities can again be postulated in advance, as there are two independent parameters to be chosen. These quantities are the rejectable level δ, and α_0, the approximate proportion of false alarms. This method of posing the problem is equivalent to the first formulation above, in which $L(0;0)$ is given and $L(0;\delta)$ must be minimized. $L(0;0)$ is assumed to be the reciprocal of α_0, but in fact this is only an approximation. If, however, there is no alternative way of setting the parameters of the cusum, any method is better than none. The method described above has great value if applied to special distributions, like the distribution of the range of a Normal sample, Binomial, Poisson, Weibull and other distributions. It would be an impossible task to compile tables of the A.R.L. for all these distributions over sufficient parameter ranges. Most of the results of this kind can be found in Johnson and Leone (1962) and Johnson (1963), (1966). For the Normal distribution we have already given the relations between h and k on the one hand and δ and α_0 on the other in section 1.4 (17).

We now turn to problems of scaling the cusum graph. There are two distinct problems here:

(1) How much room should there be on the vertical scale to prevent the cusum graph from running off the top or bottom of the paper?

(2) What should be the relation between the horizontal and vertical scales, and what is the effect on the parameters d and ϕ of the V-mask, when we change the scale factors?

Various dimensions for the vertical scale expressed in standard deviations have been recommended in the literature. The amount of room needed depends mainly on the number of observations that can be plotted on the graph horizontally, and on the size of a deviation from target that will go undetected for a number of observations which is large in proportion to the total capacity of the graph. If the observations are independent deviations of a process which is on target, the cumulative sum describes a random walk with simple properties. The distribution of the cusum after n observations is a normal distribution with zero mean and standard deviation $\sigma\sqrt{n}$. To keep such a graph on the paper with probability 0·9973 we need $3\sigma\sqrt{n}$ on both sides of the zero line. A possibly undetected slope has to be added to this.

Example. We have a graph paper for 50 observations; also, a bias of $0\cdot5\sigma$ could easily go undetected for 20 observations. The number of standard deviations needed on either side of the zero line would be $20 \times 0\cdot5 + 3\sqrt{50} = 31$. For graph paper of the usual proportions, on which the length of the vertical side is about 70 per cent of the length of the horizontal side, this implies that two standard deviations may occupy the same space on the vertical scale as one observation on the horizontal scale.

This leads us to consider the effect of this scale factor between horizontal and vertical scales on the V-mask parameters.

Let us define a scale factor w as the length taken for one observation horizontally, divided by the length for one standard deviation on the vertical scale. In the preceding example the scale factor would be 2. Obviously, the scale factor does not affect the numerical

30

cusum parameters h and k, but only the parameters of the V-mask.
The relationships between these two kinds of cusum parameters then
become

$$k = w\tan\phi \tag{32}$$

$$h = wd\tan\phi = dk. \tag{33}$$

A scale factor in the cusum graph also has its effect in other
formulas, such as (14) of section 1.4; this now becomes

$$\phi = \tan^{-1}\frac{g}{2w} \tag{34}$$

A transformation that can be useful, especially when working with
random variables which can only take positive values, such as sample
variances and Poisson variates, is the following. Supposing one were
to sample such a non-negative random variable and to plot the cumulative sum of the observations. Clearly, then, because the cusum can
only increase, both control limits will have positive slopes. The graph
will tend to run off the top of the paper very rapidly. Also, this kind of
cusum graph is difficult to interpret visually, unless the expected slope
under the null hypothesis is also shown on the graph. Fig. 9 shows
such a control chart; note that the lead distance d^-, for the test on
negative bias, is taken to the left of the last observation.

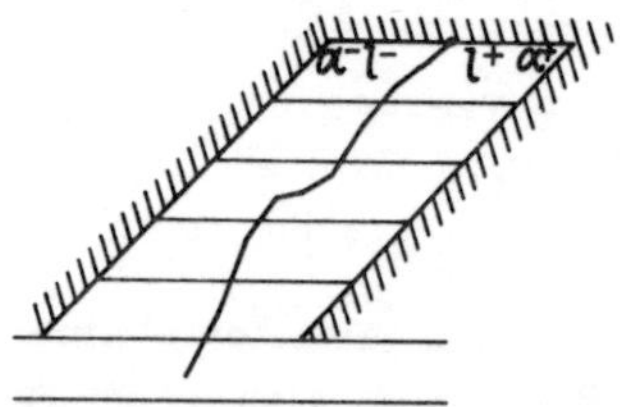

Fig. 9   Cusum mask for positive variate

Often one will want to transform such a cusum graph into a more
useful form which will stay on the paper longer and give a better idea
of the magnitude of a deviation from target, if any. This can easily be
effected by subtracting the expected sample value for a true null
hypothesis from each observation before forming the cumulative sum.
The action limits have to be changed in such a way that the cusum test
will respond at the same observations as before the transformation. If
the value subtracted from each observation is c, some simple geometry
shows that the angle α^+ and lead distance l^+ of Fig. 9 are related to

the angle ϕ^+ and lead distance d^+ of the V-mask for the transformed graph by

$$a^+ = \tan^{-1}\left\{\tan\phi^+ + \frac{c}{w}\right\} \tag{35a}$$

$$l^+ = \frac{wd^+\tan\phi^+}{w\tan\phi^+ + c} = \frac{\tan\phi^+}{\tan\alpha^+}d^+ \tag{35b}$$

and conversely

$$\phi^+ = \tan^{-1}\left\{\tan\alpha^+ - \frac{c}{w}\right\} \tag{36a}$$

$$d^+ = l^+ + \frac{l^+c}{w\tan\phi^+} = \frac{\tan\alpha^+}{\tan\phi^+}l^+. \tag{36b}$$

These relations can also be derived by realizing that the only quantity that changes in the numerical test is k^+, and using (32) and (33). For the test on negative deviations similar relations hold:

$$a^- = \tan^{-1}\left\{\tan\phi^- - \frac{c}{w}\right\} \tag{37a}$$

$$l^- = \frac{wd^-\tan\phi^-}{w\tan\phi^- - c} = \frac{\tan\phi^-}{\tan\alpha^-}d^- \tag{37b}$$

and

$$\phi^- = \tan^{-1}\left\{\tan\alpha^- + \frac{c}{w}\right\} \tag{38a}$$

$$d^- = l^- - \frac{l^-c}{w\tan\phi^-} = \frac{\tan\alpha^-}{\tan\phi^-}l^-. \tag{38b}$$

The above relations can also be used when the target value is changed, but the cusum graph continues to be plotted as before. The modified mask can then be constructed with the parameters given by the formulas.

If a negative l^- or d^- results in formulas (37b) or (38b), this means that the lead distance is to be plotted to the left of the last observation, as is indeed the case in Fig. 9.

2.2 A.R.L. tables and nomograms for tests controlling the mean of a Normal population

In this section, tables of the average run length for tests controlling the mean of a Normal distribution are given. The construction of such tables is particularly rewarding, because the number of parameters to be considered is small, since the Normal distribution shape is invariant under a change of origin. We again use θ as the symbol for the bias, expressing it in standard deviations of the observations. It is assumed that the standard deviation is known exactly in the application of the test. The standardized deviations therefore have a Normal distribution with mean θ and standard deviation unity. Subtracting or adding the cusum parameter k^+ or k^-, respectively, only has the effect of changing the mean of the accumulated values to $\theta - k^+$ or $\theta + k^-$.

It is thus possible to tabulate the average run length of cusum tests starting with a zero cusum as a function of h and $\theta - k$ only, thus covering various θ,k combinations.

In the tables that follow (Tables 5a − 5e), we give the operating characteristic, the average sample number, and the average run length of a test starting at zero. These quantities have been computed by the author on the Control Data 3600 computer and they are correct to one unit in the last decimal place given. Intermediate values may be obtained by interpolation. For interpolating the A.R.L. a satisfactory method proved to be to use the logarithms of the A.R.L. and interpolate by Everett's formula, using up to second differences with the fourth difference thrown back; see H.M.S.O. (1955).

For all practical applications the accuracy offered here is sufficient, as there are usually other factors outside the control of the statistician to influence the A.R.L. very strongly, as will be seen in section 2.4.

As is to be expected, the case of the mean of a Normal population has received relatively great attention in the publications on cusum tests. We shall make an attempt here to provide a guide to these articles in so far as they contain numerical values of the A.R.L.

The oldest and also most accurate table published is that of Barraclough and Page (1959). These are tables and charts for sequential analysis and assume that a central reference value is used. Values are given for h and the starting value Z of the test, which produce a certain given operating characteristic at zero bias, and for the bias of the alternative hypothesis (unacceptable level). As we are here only considering tests which start at $Z = 0$, only a limited number of the tabulated values can be of direct use.

The table ranges from $-1 \cdot 0$ to $+1 \cdot 0$ for $\theta - k$ in steps of $0 \cdot 25$ (except $\theta - k = 0 \cdot 0$). Values for the operating characteristic at zero deviation are $0 \cdot 999$, $0 \cdot 995$, $0 \cdot 99$ and $0 \cdot 95$. The rejectable level or θ_1 of the alternative hypothesis is given as $0 \cdot 05$, $0 \cdot 10$ and so on up to $0 \cdot 70$ in steps of $0 \cdot 10$. The average sample numbers are also given, making it possible to calculate the A.R.L., but only when $Z = 0$. The lowest and highest values of h that appear are $0 \cdot 25$ and $14 \cdot 98$.

Table 5a   Operating characteristic, average sample number, and average run length for cusum tests applied to a Normal distribution with known standard deviation ($h = 1{\cdot}3$ and $h = 1{\cdot}5$)

$\theta - k$	$h = 1{\cdot}3$			$h = 1{\cdot}5$			$\theta - k$
	P	N	L	P	N	L	
$-2{\cdot}0$	1·000	1·02	19×10^2	1·000	1·02	38×10^2	$-2{\cdot}0$
$-1{\cdot}8$	0·999	1·04	9×10^2	0·999	1·04	17×10^2	$-1{\cdot}8$
$-1{\cdot}6$	0·998	1·06·	445	0·999	1·06	800	$-1{\cdot}6$
$-1{\cdot}4$	0·995	1·09	225	0·997	1·10	380	$-1{\cdot}4$
$-1{\cdot}2$	0·990	1·14	117	0·994	1·14	186	$-1{\cdot}2$
$-1{\cdot}0$	0·981	1·20	63	0·987	1·21	94	$-1{\cdot}0$
$-0{\cdot}8$	0·964	1·28	35·4	0·974	1·31	50	$-0{\cdot}8$
$-0{\cdot}6$	0·934	1·39	20·9	0·948	1·43	27·6	$-0{\cdot}6$
$-0{\cdot}4$	0·884	1·51	13·1	0·903	1·58	16·4	$-0{\cdot}4$
$-0{\cdot}2$	0·811	1·64	8·65	0·833	1·74	10·4	$-0{\cdot}2$
$0{\cdot}0$	0·713	1·74	6·08	0·734	1·88	7·08	$0{\cdot}0$
$0{\cdot}2$	0·598	1·81	4·50	0·616	1·97	5·14	$0{\cdot}2$
$0{\cdot}4$	0·479	1·83	3·50	0·491	2·00	3·93	$0{\cdot}4$
$0{\cdot}6$	0·367	1·80	2·84	0·374	1·97	3·15	$0{\cdot}6$
$0{\cdot}8$	0·271	1·73	2·38	0·274	1·90	2·62	$0{\cdot}8$
$1{\cdot}0$	0·193	1·65	2·04	0·195	1·80	2·24	$1{\cdot}0$
$1{\cdot}2$	0·134	1·56	1·80	0·135	1·69	1·96	$1{\cdot}2$
$1{\cdot}4$	0·091	1·47	1·61	0·091	1·59	1·74	$1{\cdot}4$
$1{\cdot}6$	0·060	1·38	1·47	0·060	1·48	1·58	$1{\cdot}6$
$1{\cdot}8$	0·038	1·30	1·36	0·038	1·39	1·45	$1{\cdot}8$
$2{\cdot}0$	0·024	1·24	1·27	0·024	1·31	1·34	$2{\cdot}0$
$2{\cdot}2$	0·014	1·18	1·20	0·014	1·24	1·26	$2{\cdot}2$
$2{\cdot}4$	0·008	1·13	1·14	0·008	1·18	1·19	$2{\cdot}4$
$2{\cdot}6$	0·005	1·09	1·099	0·005	1·13	1·14	$2{\cdot}6$
$2{\cdot}8$	0·003	1·06	1·068	0·003	1·10	1·10	$2{\cdot}8$
$3{\cdot}0$	0·001	1·04	1·045	0·001	1·07	1·067	$3{\cdot}0$
$4{\cdot}0$	0·000	1·00	1·004	0·000	1·01	1·006	$4{\cdot}0$

Table 5b Operating characteristic, average sample number, and average run length for cusum tests applied to a Normal distribution with known standard deviation ($h = 2 \cdot 0$ and $h = 2 \cdot 5$)

$\theta - k$	$h = 2 \cdot 0$			$h = 2 \cdot 5$			$\theta - k$
	P	N	L	P	N	L	
$-2 \cdot 0$	$1 \cdot 000$	$1 \cdot 02$	24×10^3	$1 \cdot 000$	$1 \cdot 02$	18×10^4	$-2 \cdot 0$
$-1 \cdot 8$	$1 \cdot 000$	$1 \cdot 04$	96×10^2	$1 \cdot 000$	$1 \cdot 04$	58×10^3	$-1 \cdot 8$
$-1 \cdot 6$	$1 \cdot 000$	$1 \cdot 06$	38×10^2	$1 \cdot 000$	$1 \cdot 06$	19×10^3	$-1 \cdot 6$
$-1 \cdot 4$	$0 \cdot 999$	$1 \cdot 10$	15×10^2	$1 \cdot 000$	$1 \cdot 10$	62×10^2	$-1 \cdot 4$
$-1 \cdot 2$	$0 \cdot 998$	$1 \cdot 15$	61×10	$0 \cdot 999$	$1 \cdot 16$	21×10^2	$-1 \cdot 2$
$-1 \cdot 0$	$0 \cdot 995$	$1 \cdot 23$	259	$0 \cdot 998$	$1 \cdot 24$	716	$-1 \cdot 0$
$-0 \cdot 8$	$0 \cdot 988$	$1 \cdot 35$	114	$0 \cdot 995$	$1 \cdot 37$	262	$-0 \cdot 8$
$-0 \cdot 6$	$0 \cdot 972$	$1 \cdot 52$	54	$0 \cdot 985$	$1 \cdot 57$	104	$-0 \cdot 6$
$-0 \cdot 4$	$0 \cdot 938$	$1 \cdot 74$	$28 \cdot 0$	$0 \cdot 960$	$1 \cdot 86$	$46 \cdot 1$	$-0 \cdot 4$
$-0 \cdot 2$	$0 \cdot 875$	$1 \cdot 99$	$15 \cdot 9$	$0 \cdot 904$	$2 \cdot 22$	$23 \cdot 3$	$-0 \cdot 2$
$0 \cdot 0$	$0 \cdot 776$	$2 \cdot 24$	$10 \cdot 0$	$0 \cdot 807$	$2 \cdot 59$	$13 \cdot 4$	$0 \cdot 0$
$0 \cdot 2$	$0 \cdot 649$	$2 \cdot 41$	$6 \cdot 86$	$0 \cdot 673$	$2 \cdot 86$	$8 \cdot 73$	$0 \cdot 2$
$0 \cdot 4$	$0 \cdot 512$	$2 \cdot 47$	$5 \cdot 06$	$0 \cdot 526$	$2 \cdot 96$	$6 \cdot 24$	$0 \cdot 4$
$0 \cdot 6$	$0 \cdot 386$	$2 \cdot 43$	$3 \cdot 96$	$0 \cdot 392$	$2 \cdot 91$	$4 \cdot 79$	$0 \cdot 6$
$0 \cdot 8$	$0 \cdot 280$	$2 \cdot 33$	$3 \cdot 24$	$0 \cdot 283$	$2 \cdot 77$	$3 \cdot 87$	$0 \cdot 8$
$1 \cdot 0$	$0 \cdot 198$	$2 \cdot 20$	$2 \cdot 74$	$0 \cdot 199$	$2 \cdot 60$	$3 \cdot 25$	$1 \cdot 0$
$1 \cdot 2$	$0 \cdot 136$	$2 \cdot 05$	$2 \cdot 38$	$0 \cdot 137$	$2 \cdot 42$	$2 \cdot 80$	$1 \cdot 2$
$1 \cdot 4$	$0 \cdot 092$	$1 \cdot 91$	$2 \cdot 10$	$0 \cdot 092$	$2 \cdot 25$	$2 \cdot 47$	$1 \cdot 4$
$1 \cdot 6$	$0 \cdot 060$	$1 \cdot 78$	$1 \cdot 89$	$0 \cdot 060$	$2 \cdot 09$	$2 \cdot 22$	$1 \cdot 6$
$1 \cdot 8$	$0 \cdot 038$	$1 \cdot 65$	$1 \cdot 72$	$0 \cdot 038$	$1 \cdot 94$	$2 \cdot 02$	$1 \cdot 8$
$2 \cdot 0$	$0 \cdot 024$	$1 \cdot 54$	$1 \cdot 58$	$0 \cdot 024$	$1 \cdot 81$	$1 \cdot 85$	$2 \cdot 0$
$2 \cdot 2$	$0 \cdot 014$	$1 \cdot 44$	$1 \cdot 46$	$0 \cdot 014$	$1 \cdot 69$	$1 \cdot 71$	$2 \cdot 2$
$2 \cdot 4$	$0 \cdot 008$	$1 \cdot 36$	$1 \cdot 37$	$0 \cdot 008$	$1 \cdot 58$	$1 \cdot 59$	$2 \cdot 4$
$2 \cdot 6$	$0 \cdot 005$	$1 \cdot 28$	$1 \cdot 28$	$0 \cdot 005$	$1 \cdot 48$	$1 \cdot 49$	$2 \cdot 6$
$2 \cdot 8$	$0 \cdot 003$	$1 \cdot 21$	$1 \cdot 22$	$0 \cdot 002$	$1 \cdot 39$	$1 \cdot 40$	$2 \cdot 8$
$3 \cdot 0$	$0 \cdot 001$	$1 \cdot 16$	$1 \cdot 16$	$0 \cdot 001$	$1 \cdot 31$	$1 \cdot 32$	$3 \cdot 0$
$4 \cdot 0$	$0 \cdot 000$	$1 \cdot 02$	$1 \cdot 023$	$0 \cdot 000$	$1 \cdot 07$	$1 \cdot 067$	$4 \cdot 0$

Table 5c Operating characteristic, average sample number, and average
run length for cusum tests applied to a Normal distribution with known
standard deviation ($h = 3.0$ and $h = 4.0$)

$\theta - k$	$h = 3.0$			$h = 4.0$			$\theta - k$
	P	N	L	P	N	L	
-2.0	1.000	1.02	$>10^7$	1.000	1.02	$>10^8$	-2.0
-1.8	1.000	1.04	$>10^6$	1.000	1.04	$>10^7$	-1.8
-1.6	1.000	1.06	1×10^5	1.000	1.06	2×10^6	-1.6
-1.4	1.000	1.10	26×10^3	1.000	1.10	42×10^4	-1.4
-1.2	1.000	1.16	70×10^2	1.000	1.16	76×10^3	-1.2
-1.0	0.999	1.25	20×10^2	1.000	1.25	14×10^3	-1.0
-0.8	0.998	1.39	590	0.999	1.40	29×10^2	-0.8
-0.6	0.992	1.61	195	0.998	1.65	660	-0.6
-0.4	0.974	1.95	73.6	0.988	2.08	178	-0.4
-0.2	0.926	2.43	32.8	0.954	2.78	60.3	-0.2
0.0	0.830	2.94	17.3	0.863	3.65	26.6	0.0
0.2	0.689	3.32	10.7	0.711	4.30	14.9	0.2
0.4	0.535	3.46	7.44	0.544	4.51	9.88	0.4
0.6	0.396	3.39	5.62	0.399	4.37	7.28	0.6
0.8	0.284	3.22	4.49	0.285	4.11	5.74	0.8
1.0	0.199	3.00	3.75	0.199	3.80	4.75	1.0
1.2	0.137	2.78	3.22	0.137	3.50	4.06	1.2
1.4	0.092	2.58	2.84	0.092	3.22	3.55	1.4
1.6	0.060	2.39	2.54	0.060	2.97	3.16	1.6
1.8	0.038	2.22	2.31	0.038	2.75	2.86	1.8
2.0	0.024	2.07	2.12	0.024	2.56	2.62	2.0
2.2	0.014	1.94	1.96	0.014	2.39	2.42	2.2
2.4	0.008	1.82	1.83	0.008	2.24	2.26	2.4
2.6	0.005	1.71	1.72	0.005	2.12	2.13	2.6
2.8	0.003	1.61	1.61	0.003	2.01	2.02	2.8
3.0	0.001	1.52	1.52	0.001	1.92	1.92	3.0
4.0	0.000	1.16	1.16	0.000	1.50	1.50	4.0

Table 5d Operating characteristic, average sample number, and average run length for cusum tests applied to a Normal distribution with known standard deviation ($h = 5 \cdot 0$ and $h = 6 \cdot 0$)

$\theta - k$	$h = 5 \cdot 0$			$h = 6 \cdot 0$			$\theta - k$
	P	N	L	P	N	L	
$-1 \cdot 0$	1·000	1·25	$> 10^5$	1·000	1·25	$> 7 \times 10^5$	$-1 \cdot 0$
$-0 \cdot 8$	1·000	1·40	14×10^3	1·000	1·40	7×10^4	$-0 \cdot 8$
$-0 \cdot 6$	0·999	1·66	22×10^2	1·000	1·66	74×10^2	$-0 \cdot 6$
$-0 \cdot 4$	0·995	2·15	414	0·998	2·19	940	$-0 \cdot 4$
$-0 \cdot 2$	0·971	3·06	104	0·981	3·28	171	$-0 \cdot 2$
$0 \cdot 0$	0·886	4·36	38·1	0·902	5·06	51·6	$0 \cdot 0$
$0 \cdot 2$	0·725	5·34	19·4	0·733	6·42	24·0	$0 \cdot 2$
$0 \cdot 4$	0·548	5·58	12·4	0·550	6·69	14·9	$0 \cdot 4$
$0 \cdot 6$	0·400	5·37	8·94	0·400	6·36	10·6	$0 \cdot 6$
$0 \cdot 8$	0·285	5·00	6·99	0·285	5·89	8·24	$0 \cdot 8$
$1 \cdot 0$	0·199	4·60	5·75	0·199	5·40	6·75	$1 \cdot 0$
$1 \cdot 2$	0·137	4·22	4·89	0·137	4·94	5·72	$1 \cdot 2$
$1 \cdot 4$	0·092	3·87	4·26	0·092	4·52	4·98	$1 \cdot 4$
$1 \cdot 6$	0·060	3·56	3·79	0·060	4·15	4·41	$1 \cdot 6$
$1 \cdot 8$	0·038	3·28	3·41	0·038	3·82	3·97	$1 \cdot 8$
$2 \cdot 0$	0·024	3·04	3·11	0·024	3·53	3·62	$2 \cdot 0$
$2 \cdot 2$	0·014	2·83	2·87	0·014	3·28	3·33	$2 \cdot 2$
$2 \cdot 4$	0·008	2·64	2·66	0·008	3·06	3·08	$2 \cdot 4$
$2 \cdot 6$	0·005	2·48	2·49	0·005	2·86	2·88	$2 \cdot 6$
$2 \cdot 8$	0·003	2·34	2·35	0·003	2·69	2·70	$2 \cdot 8$
$3 \cdot 0$	0·001	2·22	2·23	0·001	2·54	2·54	$3 \cdot 0$
$4 \cdot 0$	0·000	1·86	1·86	0·000	2·06	2·06	$4 \cdot 0$

Table 5e   Operating characteristic, average sample number, and average run length for cusum tests applied to a Normal distribution with known standard deviation ($h = 8 \cdot 0$ and $h = 10 \cdot 0$)

$\theta - k$	$h = 8 \cdot 0$			$h = 10 \cdot 0$			$\theta - k$
	P	N	L	P	N	L	
$-1 \cdot 0$	$1 \cdot 000$	$1 \cdot 25$	$> 10^7$	$1 \cdot 000$	$1 \cdot 25$	$> 10^7$	$-1 \cdot 0$
$-0 \cdot 8$	$1 \cdot 000$	$1 \cdot 40$	2×10^5	$1 \cdot 000$	$1 \cdot 40$	$> 10^6$	$-0 \cdot 8$
$-0 \cdot 6$	$1 \cdot 000$	$1 \cdot 67$	82×10^3	$1 \cdot 000$	$1 \cdot 67$	$\approx 10^6$	$-0 \cdot 6$
$-0 \cdot 4$	$1 \cdot 000$	$2 \cdot 22$	47×10^2	$1 \cdot 000$	$2 \cdot 23$	2×10^5	$-0 \cdot 4$
$-0 \cdot 2$	$0 \cdot 992$	$3 \cdot 59$	428	$0 \cdot 995$	$3 \cdot 78$	820	$-0 \cdot 2$
$0 \cdot 0$	$0 \cdot 923$	$6 \cdot 48$	$84 \cdot 6$	$0 \cdot 937$	$7 \cdot 93$	126	$0 \cdot 0$
$0 \cdot 2$	$0 \cdot 742$	$8 \cdot 70$	$33 \cdot 6$	$0 \cdot 745$	$11 \cdot 01$	$43 \cdot 2$	$0 \cdot 2$
$0 \cdot 4$	$0 \cdot 552$	$8 \cdot 91$	$19 \cdot 9$	$0 \cdot 552$	$11 \cdot 16$	$24 \cdot 9$	$0 \cdot 4$
$0 \cdot 6$	$0 \cdot 400$	$8 \cdot 36$	$13 \cdot 9$	$0 \cdot 400$	$10 \cdot 38$	$17 \cdot 3$	$0 \cdot 6$
$0 \cdot 8$	$0 \cdot 285$	$7 \cdot 68$	$10 \cdot 7$	$0 \cdot 285$	$9 \cdot 46$	$13 \cdot 2$	$0 \cdot 8$
$1 \cdot 0$	$0 \cdot 199$	$7 \cdot 00$	$8 \cdot 75$	$0 \cdot 199$	$8 \cdot 60$	$10 \cdot 7$	$1 \cdot 0$
$1 \cdot 2$	$0 \cdot 137$	$6 \cdot 38$	$7 \cdot 39$	$0 \cdot 137$	$7 \cdot 82$	$9 \cdot 06$	$1 \cdot 2$
$1 \cdot 4$	$0 \cdot 092$	$5 \cdot 82$	$6 \cdot 41$	$0 \cdot 092$	$7 \cdot 12$	$7 \cdot 84$	$1 \cdot 4$
$1 \cdot 6$	$0 \cdot 060$	$5 \cdot 32$	$5 \cdot 66$	$0 \cdot 060$	$6 \cdot 50$	$6 \cdot 91$	$1 \cdot 6$
$1 \cdot 8$	$0 \cdot 038$	$4 \cdot 91$	$5 \cdot 11$	$0 \cdot 038$	$5 \cdot 96$	$6 \cdot 19$	$1 \cdot 8$
$2 \cdot 0$	$0 \cdot 024$	$4 \cdot 51$	$4 \cdot 62$	$0 \cdot 024$	$5 \cdot 48$	$5 \cdot 62$	$2 \cdot 0$
$2 \cdot 2$	$0 \cdot 014$	$4 \cdot 19$	$4 \cdot 24$	$0 \cdot 014$	$5 \cdot 07$	$5 \cdot 14$	$2 \cdot 2$
$2 \cdot 4$	$0 \cdot 008$	$3 \cdot 88$	$3 \cdot 92$	$0 \cdot 008$	$4 \cdot 71$	$4 \cdot 75$	$2 \cdot 4$
$2 \cdot 6$	$0 \cdot 005$	$3 \cdot 63$	$3 \cdot 65$	$0 \cdot 005$	$4 \cdot 40$	$4 \cdot 42$	$2 \cdot 6$
$2 \cdot 8$	$0 \cdot 003$	$3 \cdot 41$	$3 \cdot 42$	$0 \cdot 003$	$4 \cdot 12$	$4 \cdot 14$	$2 \cdot 8$
$3 \cdot 0$	$0 \cdot 001$	$3 \cdot 22$	$3 \cdot 23$	$0 \cdot 001$	$3 \cdot 88$	$3 \cdot 89$	$3 \cdot 0$
$4 \cdot 0$	$0 \cdot 000$	$2 \cdot 51$	$2 \cdot 51$	$0 \cdot 000$	$3 \cdot 05$	$3 \cdot 05$	$4 \cdot 0$

This table is primarily of interest for the design of sequential acceptance schemes; it is mentioned here because it is the only accurate table available in the published literature and certain entries in it have been used repeatedly to check the accuracy of approximations.

Ewan and Kemp (1960) published a table of A.R.L. figures calculated by an approximate method (see also section 3.3) and covering the range $h = 2 \cdot 0\ (1 \cdot 0)\ 5 \cdot 0$; for each of these values of h the A.R.L. is given for about 15 values of $\theta - k$. The errors in this table range from zero to about 5 per cent of the true A.R.L. A nomogram of the A.R.L., which is rather useful, is included.

Goldsmith and Whitfield (1961) give a rather more extensive set of graphs of the A.R.L. as a function of θ for 16 different combinations of k and h. Because the graphs are arranged by d and ϕ, the parameters of the graphical test, some overlap occurs on certain parts of the graphs. For example, curves IV and V both have $h = 1 \cdot 6$ (note that a scale factor $w = 2$ is used throughout). Curves XII and XIV both have $h = 4 \cdot 0$.

Points on the graphs were found for $\theta = 0 \cdot 0,\ 0 \cdot 5,\ 1 \cdot 0,\ 2 \cdot 0,$ and $3 \cdot 0$. The results were obtained by Monte Carlo methods on a digital computer. This technique, which was also used initially by the author, is further discussed in section 3.6. This method is usually very expensive when relatively high accuracy is required, because halving the error means quadrupling the amount of computation. It is reported in the article that the coefficient of variation in the figures found never exceeds 10 per cent, which is regarded by the author as sufficiently accurate for most practical purposes.

A very interesting result in this paper is that a further simulation, in which the successive observations were serially correlated, showed that the A.R.L. is not too sensitive to the degree of serial correlation for a fair range of the serial correlation coefficient. This kind of robustness is always very encouraging to the prospective user of a method.

Another note about interpolating the A.R.L. for fairly high values of $\theta - k$ is in place here. Various authors mention that for these values of $\theta - k$ the A.R.L. may be approximated by a formula of the form

$$L = R + \frac{h}{\theta - k}. \tag{39}$$

For the constant R, different values are cited. They range from $0 \cdot 66$ to $1 \cdot 0$. There are good reasons for a relation of this kind to hold true, as will be shown in section 3.5. Also the value of R is not a constant, but may be expected to vary with $\theta - k$, whence the different empirical values quoted in the literature; one value may be the best for a certain range of $\theta - k$ and another value for a different range.

In section 3.5 a table is given of $\hat{R} = L - h/(\theta - k)$ for the range of parameters for which we have tables in this section. This property may be exploited for interpolation in the tables of this section, especially if neither h nor $\theta - k$ appear in the table entries. One can then estimate R from the four surrounding values and use (39). If R appears to change appreciably between the values considered, any simple (linear) method of interpolation in a two-way table will generally give results accurate to within two units of the last decimal place of the A.R.L.

We do not give any $L(0;\theta)$ graphs here; one such graph has already been shown in section 1.5, Fig. 8. Some graphs for two-sided tests are given in section 2.4. When using two-sided cusum tests, one should always calculate the A.R.L. from the A.R.L. for each test using (21). Often one is unpleasantly surprised because (for a symmetrical test) the $L(0;0)$ halves, whereas the change for large positive or negative θ is negligible. For

$$L = \frac{L^+ L^-}{L^+ + L^-} = \frac{L^+}{1 + L^+/L^-} \approx L^+, \quad \text{if} \quad L^- \gg L^+, \tag{40}$$

and similarly for large L^+.

In Table 6 an example is given of the calculation of the A.R.L. of an asymmetrical two-sided cusum test. All these values of L^+ and L^- can be read off directly from Table 5c.

Table 6   Example of the calculation of the A.R.L. of a two-sided cusum test

$$h^+ = h^- = 3{\cdot}0, \quad k^+ = 0{\cdot}4, \quad k^- = 0{\cdot}8$$

θ	L^-	L^+	L
$-2{\cdot}0$	$3{\cdot}22$	$>10^7$	$3{\cdot}22$
$-1{\cdot}6$	$4{\cdot}49$	10^7	$4{\cdot}49$
$-1{\cdot}2$	$7{\cdot}44$	10^5	$7{\cdot}44$
$-0{\cdot}8$	$17{\cdot}3$	7000	$17{\cdot}3$
$-0{\cdot}4$	$73{\cdot}6$	590	$65{\cdot}4$
$0{\cdot}0$	590	$73{\cdot}6$	$65{\cdot}4$
$0{\cdot}4$	7000	$17{\cdot}3$	$17{\cdot}3$
$0{\cdot}8$	10^5	$7{\cdot}44$	$7{\cdot}44$
$1{\cdot}2$	10^7	$4{\cdot}49$	$4{\cdot}49$
$1{\cdot}6$	$>10^7$	$3{\cdot}22$	$3{\cdot}22$
$2{\cdot}0$	$>10^7$	$2{\cdot}54$	$2{\cdot}54$

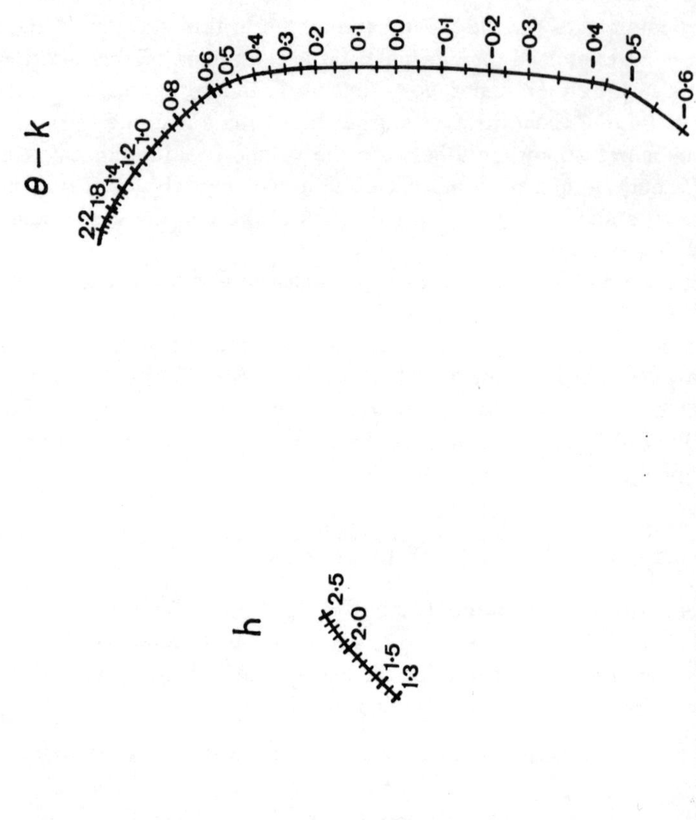

Fig. 10 Nomogram of A.R.L. of cusum tests for Normal distribution

Fig. 10 is a nomogram of the A.R.L. for one-sided tests for a range of h from 1·3 to 2·5 and $\theta - k$ from −0·6 to 2·2. This nomogram was constructed by several iterations, using the method described in Lyle (1954), on tabulated values of Table 5.

A possibly more useful graph for deciding between alternative choices of cusum test parameters is given as Fig. 11. In this diagram

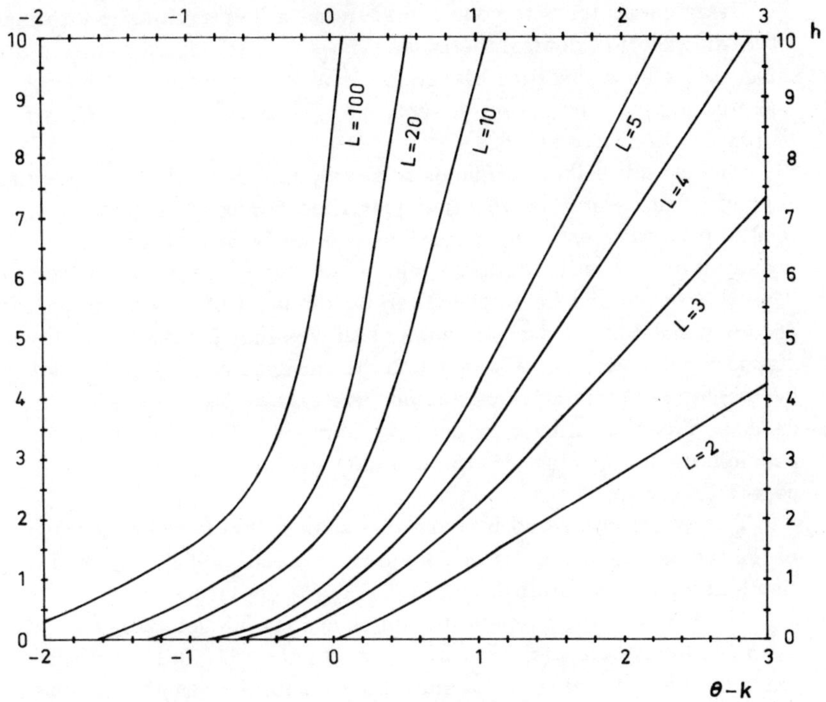

Fig. 11 $L(h; \theta-k)$ plots useful for finding "best tests" (Normal distribution)

the loci of equal A.R.L. are shown in the parameter space of h and $\theta - k$. Note that for low A.R.L., the loci are practically straight lines through the origin, which supports approximation (39). R corresponds to the slope of the lines in Fig. 11.

An optimum choice of cusum parameters can to some degree of approximation be made by using the method illustrated in section 2.1 (Table 4) graphically. A horizontal length θ_1 is moved along the locus of the chosen $L(0; \theta_1)$; the location of the left-hand end of this length

42

indicates whether a maximum $L(0;0)$ has been reached or not. Starting
off with given $L(0;0)$, minimizing $L(0;\theta_1)$ is done in a similar way by
locating the length θ_1 to the right of the chosen locus. From Fig. 11
we also have confirmation of the relatively flat extremum that is obtained.

2.3   Cusum tests controlling dispersion

Using cusum tests to control dispersion is in principle no different
from using them to control the mean. The simplest method to arrive at
cusum tests for controlling dispersion is to use Johnson's method of
constructing tests as shown in section 1.4 (see also the second part
of Johnson and Leone (1962)).

One can use either variances or sample ranges as the test quantity.
As both of these are non-negative quantities, a plot of cumulative
results from samples will always rise. Naturally, it is possible to
construct a mask with two limbs with a positive slope, as is in fact
done in Johnson and Leone (1962). From the point of view of graphical
accuracy and ease of use, the author believes that it is better to use a
standardized chart, i.e. one in which the variance or range of the null
hypothesis is subtracted from the observations before calculating the
cusums. Transformation from one type of chart to the other is always
possible using formulas (35) through (38), given in section 2.1 (see
page 31).

The reader interested in using this method is referred to the use
of the tables and nomogram in Johnson and Leone (1962). In conjunc-
tion with these, transformations (35) − (38) can be used.

Another very approximate, but quick and useful method is the following. Standardize each variance by subtracting its null-hypothesis
value and then dividing by its standard error for the sample size used.
Treat these standardized values as if they were observations from a
Normal population and use Table 5 to choose parameters for the cusum
test. The standard error to be used is $\sigma_0^2 \sqrt{2/n}$, if σ_0^2 is the variance
under the null hypothesis and n the sample size used.

This procedure may seem very rough, but as will be shown in the
next section, the A.R.L. will not be very different from the one indicated by the Normal approximation for not too small a sample (i.e. not
less than about 20 observations per sample).

Even better results may be obtained by using the logarithmic transformation of the sample variance − compare Bartlett and Kendall (1946).
As the transformed variables approach the Normal distribution more
quickly than the untransformed ones, hence the A.R.L. tables are more
correct.

2.4 Robustness of the A.R.L. with respect to distribution shape

In section 2.2 we gave a number of tables for the A.R.L. of cusum tests. All of these are based on a normal distribution of observations. It is therefore important to know something about the sensitivity of the results of those tables with respect to distribution shape, in order to enable one to decide on the useful range of applications for Tables 5.

Two further assumptions have been made implicitly: (1) the standard deviation is assumed to be known exactly, and (2) successive observations are supposed to be statistically independent. We take each of these three cases in turn in this section to see what information is available about the sensitivity of the A.R.L. with respect to these assumptions.

To test the sensitivity of the A.R.L. with respect to distribution shape, a few selected A.R.L. values were calculated for the negative exponential distribution as the extreme, and for various less skew Gamma distributions. Note that the negative exponential itself is a Gamma distribution with shape parameter 1. As the Gamma distribution tends to a Normal distribution as its shape parameter approaches infinity, one expects the A.R.L. of a particular cusum test to get closer to that for a Normal distribution as the shape parameter of the Gamma distribution increases. This effect is seen in Table 7 below for $h = 3$. In all cases distributions have been standardized, i.e. transformed into a distribution with zero mean and unit standard deviation.

Table 7 The A.R.L. of cusum tests with $h = 3$ for various Gamma distributions of observations

$\theta - k$	Gamma parameter					Normal
	1	2	3	4	5	
−1·0	108	152	191	226	257	2000
−0·5	50·6	57·1	61·7	65·2	68·0	118
0·0	19·7	18·7	18·3	18·1	18·0	17·3
0·5	7·51	7·14	7·00	6·90	6·84	6·41
1·0	3·97	3·96	3·92	3·90	3·88	3·75

We see that for negative $\theta - k$ there is a considerable difference in the A.R.L. for low shape parameter. For increasing $\theta - k$ and also for increasing shape parameter of the Gamma distribution there is quick convergence towards the A.R.L. which is obtained with a Normal distribution.

44

Admittedly, this is a very restricted example and it would be useful to conduct a more detailed investigation into the limits of possible use of the tables for the A.R.L. as given.

The remarks made in the preceding section are now justified to a certain extent, since the distribution shape of the sample variance of samples of size N from a Normal distribution is that of a Gamma distribution with shape parameter $(N-1)/2$. So for sample size $N = 11$, the distribution of sample variances has the shape of a Gamma distribution with parameter 5, which yields results close to those of the Normal distribution.

Secondly, the fact that the standard deviation is not known exactly, as is tacitly assumed in the cusum test procedure, is much more serious than slight deviations from normality. Usually one estimates the standard deviation from the same observations as those used in the test, and errors of the order of 10 to 20 per cent are not uncommon.

The effects of over- and underestimation of the standard deviation by a factor are to multiply both cusum parameters effectively by that same factor. In Fig. 12 the A.R.L. curve is given for a two-sided cusum test with parameters $h = 2\cdot5$, $k = 0\cdot5$, together with the A.R.L.

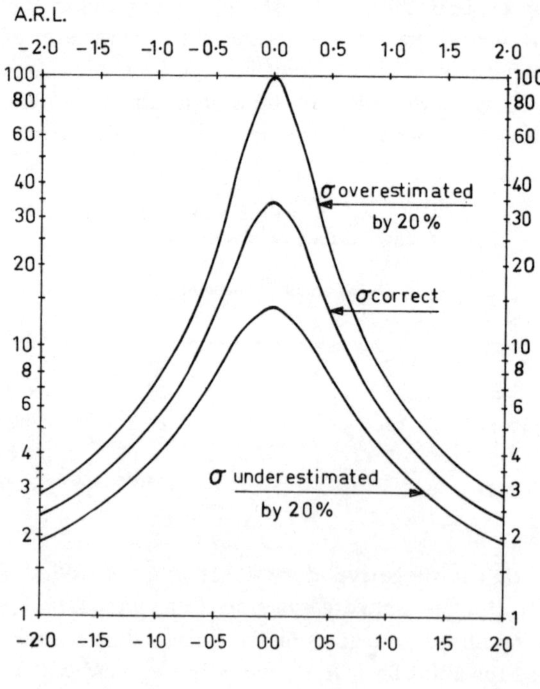

Fig 12　Effect of over- or underestimating σ for cusum test ($h = 2\cdot5$, $k = 0\cdot5$)

curves for the cases in which the standard deviation is too high or too low by 20 per cent. These results can be read off directly from Table 5. The effects are seen to be rather dramatic; unfortunately there appears to be little one can do about it. One cannot even use the t-distribution, as the standard deviations used for successive cusum updates will usually be calculated from increasing samples and therefore successive estimates of the standard deviation will be serially correlated. This may be a field in which there is some point in getting more information about what is happening by Monte Carlo methods (see section 3.6). Very much depends on whether the standard deviation is updated at every observation or not. If it is not, the A.R.L. is given by one of the curves of the class pictured in Fig. 12; short of knowing it, one can by calculating a weighted A.R.L. curve find the expected A.R.L.; the weights are given by the prior distribution of the estimates of the standard deviation.

The effect of serial correlation between successive observations, which is extremely difficult to study analytically, has been investigated by Goldsmith and Whitfield (1961) through Monte Carlo methods. It appears that for high $\theta - k$ (or low A.R.L.) there is little dependence of the A.R.L. on the parameter of a first-order autoregressive process (Markov process), except when the parameter comes close to 1 (when there is complete serial dependence). For higher A.R.L., the effect is to decrease the A.R.L. when the first serial correlation coefficient is positive, and to increase the A.R.L. when this coefficient is negative.

2.5 Control of forecast-errors of routine forecasting systems

Cusum tests are very suited for use as a "tracking signal" in routine forecasting systems. Especially when it would be very costly to have more frequent updating of the forecasts considered, a steep A.R.L. (θ)-curve is desirable in order to signal incorrect forecasts as promptly as possible, even with relatively small θ.

The introduction of the numerical cusum test version into a computerized forecasting system does not cause any difficulties, as the calculation is simple and the extra storage requirement modest. For each item, it is necessary to store the standard deviation and a positive and negative sum. This is under the assumption that the same cusum parameters are used for a group of articles, as is usually the case. On the other hand, the graphical version of the cusum test is a useful aid to forecasting in those cases where there is not too large a number of items to be monitored. The great graphic advantages, which contribute so much to the understanding of the processes at work, are obvious in this field.

We have already mentioned the application of cusum techniques as a forecasting tool in section 1.2. If changing the forecast has the effect of increasing costs, e.g. because a production plan has to be recalculated, cusum tests can be of immense value.

There is a complication in using the standard A.R.L. tables when the quantities under review are the forecast errors from an adaptive predictor. This is caused by the fact that the average deviation from target is not a constant, but is continually being driven back to zero by the predictor, if it is a good one. So it may happen that an average error θ exists just after a step change of average demand, but before a signal is triggered by the cusum test the average error has been reduced to some small fraction of θ after a number of observations. This has the effect of producing longer runs, and hence of increasing the A.R.L. as compared with the standard tables.

The magnitude of this effect can be determined accurately with some computing effort. We shall give some results below. There is an additional effect, however, which is not dealt with so easily. This is an effect brought about by the serial correlation in the forecast-errors of adaptive predictors and is especially apparent for high A.R.L. (see also the preceding section on Goldsmith and Whitfield's (1961) findings). The nature of the autocorrelation of forecast-errors of single exponential smoothing is discussed in Brown (1963); it is slightly more complicated than that of the Markov process studied by Goldsmith and Whitfield (1961) and can best be compared with a Markov process with negative parameter. The effect is to increase the A.R.L.

A heuristic explanation for the effect is the following: when the A.R.L. is high, the cusum needs a fair number of observations to build up to the trigger level. However, the fact that it builds up means that the forecast-errors in the section of the time-series under consideration are biased in one direction. This in turn will cause the adaptive predictor to correct the forecast so as to reduce this same bias, thus inhibiting further building up of the cusum. Hence the overall effect is to produce a much higher A.R.L. for low θ than standard theory predicts.

The numerical evaluation of the A.R.L. under conditions of serial correlation is hampered by grave difficulties. Therefore the author has carried out some simulations to obtain an idea of the approximate magnitude of the effect. The accuracy of the results is to within about 2 per cent. It has not been possible to ascertain the magnitude of the effect at high values of the A.R.L., e.g. at a value of the A.R.L. of 1000, because of the prohibitive amount of computing time that would have been required (simulation time being proportional to the A.R.L. for a given precision).

The two effects on the A.R.L. discussed above, and illustrated numerically below, also occur when the cusum test is used for quality control on articles produced by equipment which has automatic controls on it to keep the process results near the set point. In both types of application one finds that the cusum test and the automatic controls (or predictor) are complementary. If the one is slow – or has a large time constant – the other will react faster. It depends very much on the actual situation one is in, where the most weight is laid. In any case it can be misleading to design the two parts of a system separately.

In Fig. 13 and 14 the A.R.L. (θ)-graphs are given for two single-sided cusum tests. In each graph not only the A.R.L. as given by the standard tables is shown, but also the A.R.L. for tests on forecast-errors of single exponential smoothing for a few values of the smoothing parameter α (note that this is only the first effect caused by changing mean forecast-error).

Single exponential smoothing has been chosen for the example, because it is the simplest known adaptive predictor and also one which is in fairly wide use. The effect as shown is quite general, however, to any kind of adaptive predictor formula. References to selected papers on adaptive predictors are marked in the bibliography at the end of this monograph by the letter E.

For single exponential smoothing, if the mean forecast-error is θ at time zero and the mean of the time-series does not change, the mean forecast-error is a monotonically decreasing function of the observation number:

$$\theta_t = (1 - \alpha)^t \, \theta_0 \qquad t \geqslant 0.$$

The data on which Fig. 13 and 14 are based have been computed using this relationship and calculating the A.R.L. backwards, in much the same way in which functional equations of dynamic programming are often solved. The computation is started off with a small $\theta_t = \epsilon$, small enough to give only a slight difference in the A.R.L. with $\theta = 0$. When the A.R.L. has been calculated for that value, the calculation is made for $\theta = \epsilon/(1 - \alpha)$, which is the mean forecast-error before the period in which $\theta = \epsilon$. In this calculation the results of the previous step are used. This procedure is repeated until the maximum value of θ has been reached, which is still of interest.

The effect shown in Fig. 13 and 14 would describe the situation completely, if only the changing mean forecast-error of a certain pattern had to be considered. There is however – at least when using single exponential smoothing – the effect of serial correlation of the forecast-errors (or more correctly: of the difference between the

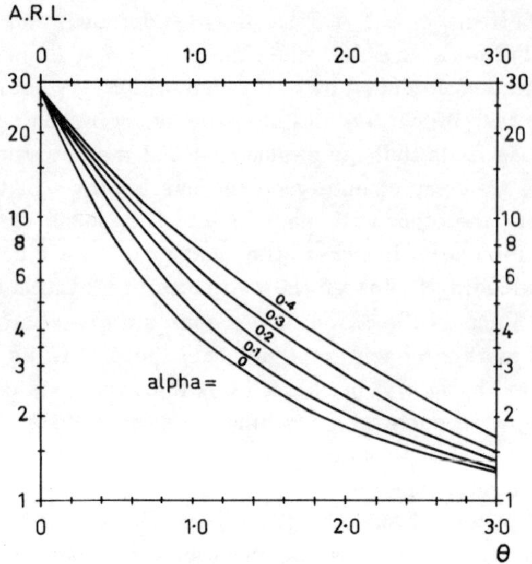

Fig. 13 A.R.L. of test ($h = 2$, $k = 0\cdot4$) after step change with exponential smoothing

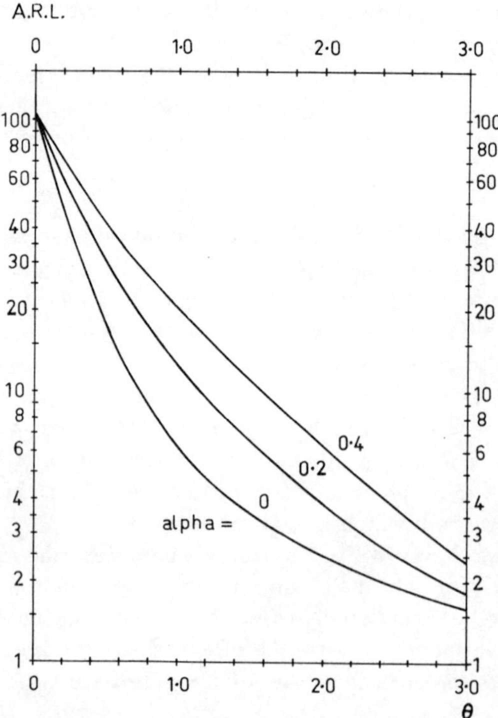

Fig. 14 A.R.L. of test ($h = 2\cdot5$, $k = 0\cdot6$) after step change with exponential smoothing

forecast-errors and their expectation). In the predictor these effects
are additive because the predictor is a linear filter. In their effect on
the A.R.L. of the cusum they may not be additive because the cusum
test is non-linear. The best we can do, however, to estimate the
effect is to assume that it is approximately additive to the first
effect, and hence to assume that a fixed percentage of the A.R.L.
is added to it for a particular α and theoretical A.R.L.

To estimate these percentages, simulations have been done for
$\theta = 0$, in which case only the serial correlation effect is present. The
results are given in Table 8. The results seem to be fairly consistently
a function of the A.R.L. rather than varying considerably for different
test parameters. Because of this, one may feel justified in adding a
certain percentage to the A.R.L., which percentage is only a function
of the A.R.L. shown in the standard tables and the smoothing constant.
For $\alpha = 0.2$ this percentage would be about 45 per cent at an A.R.L. of
25 and about 60 per cent at an A.R.L. of 50.

Table 8   A.R.L. of cusum tests on forecast-errors of single exponential
smoothing for $\theta = 0$ (Normal distribution)

α	$k = 0.4$ $h = 2.5$	0.6 2.0	0.4 1.5	0.2 2.5	0.2 1.3	0.4 1.3	0.2 2.0	0.4 2.0
0	46.1	54.3	16.4	23.3	8.65	13.1	15.9	28.0
0.1	61.0	66.4	18.4	28.8	9.3	14.3	18.5	34.0
0.2	77.6	78.6	20.1	34.4	10.0	15.3	20.7	39.8
0.3	99.2	92.4	21.7	41.0	10.5	16.6	22.8	44.8
0.4			23.1	50.4	11.0	17.4	25.5	52.2

2.6   Difficulties in routine forecasting: the top ten method

The usual picture of a system of adaptive forecasting and tracking
signals in industry is that of the presence of thousands of items, sub-
divided into article groups of a few hundred items. For each article
group there is an article group manager, with responsibility for the forecasts for these items and a knowledge of their market possibilities.

A load of routine work has been taken off the shoulders of the
article group manager by the automatic—maybe even computerized—
forecasting system. The manager can therefore devote his time to those
cases which merit his special attention. For the items in the automatic
forecasting system, the tracking signal acts as the filter to sort out
such cases.

If considering the items for which the tracking signal has triggered a query takes up a major proportion of the manager's time, a fairly constant number of triggering signals per cycle is desirable. A variable load can only be accommodated if that load is light on average.

This raises a problem in setting the cusum parameters. One tries to achieve a given average number of triggered signals every computation-cycle. To this end one must consider: the distribution of θ between types, variations in the estimated standard deviations, and the smoothing constant used. If all these are known exactly, it is in principle possible to select the cusum parameters in such a way as to obtain the desired average number of signals.

Not only are these quantities not known exactly, but also this will only bring the *average* number of signals within our control. There will still be considerable variations from period to period, roughly of a Binomial type (assuming an equal probability for each item to trigger at any time, this probability being the inverse of the average A.R.L.).

A simple and effective solution to this problem is to choose only one parameter for the cusum calculation, namely k, and to rank the resulting cusums during each period in decreasing absolute magnitude. The n largest absolute cusums are then taken as out of control. N is the desired number of signals per period.

If some items are more important than others, one can group all items into classes of a certain width, rank the cusums for each of these classes separately, and take the top $n_1, n_2, \ldots$, respectively.

The only apparent disadvantage of a system such as that described is that there is no indication of a disaster, such as the bottom dropping out of the whole market, or a segment of it. In the traditional system such a situation would be apparent from an extraordinarily high number of signals.

This difficulty can be circumvented by imposing limits on the nth highest cusum value. If these limits are the lower limit l and the upper limit u, the decision rule works as follows: after ranking the cusum according to absolute decreasing order, take out the top n for inspection; if the lowest absolute cusum of the top n is higher than the limit u, signal all cusums greater than u; on the other hand, if it is less than the limit l, only signal for the cusums greater than l.

In this system, with $u \geqslant l$, one normally obtains exactly n signals, but possibly more when there is a shift upwards in the distribution of absolute cusums and also possibly less when there is a shift downwards of the whole distribution. The fact of one's obtaining a number of signals different from n is in itself an extra indication of there being something unusual.

Two special variations of this rule are worth mentioning: (1) $l = u$;

this takes us back to the ordinary cusum test with $h = l = u$; (2) $l = 0$, $u = \infty$; in this case, we always get exactly n signals.

The choice of l and u is critical in determining the performance of this rule. One would like to choose them in such a way that in normal conditions the nth highest absolute cusum is between them, but also such that a shift of the cusum distribution is readily noticed. The theory of order statistics yields a simple solution here. The nth highest absolute cusum is the $(N-n+1)$st order statistic of the absolute cusum values.

The percentage points of an order statistic can be expressed in percentage points of the parent distribution (here the distribution of absolute values of the cusums). For a particular percentage point of the parent distribution for which the probability of being exceeded is p, the probability of obtaining n out of a sample of size N above this point is

$$\frac{N!}{(N-n)! \, n!} \, p^n (l-p)^{N-n},$$

a Binomial probability.

To obtain the percentile of the parent distribution which will give a desired percentile of the order statistic, an iterative procedure or a good table of the incomplete Beta function is needed (Harter (1964)). In Table 9 some values are given which are not easily obtained from published tables and are of interest for this particular application. They have been computed by an iterative procedure. It can be seen that for a large enough group of articles N, the distribution of the order statistic is very sensitive. This means that if the lower and upper limits for the nth highest cusum have been chosen for the null-hypothesis cusum distribution, a slightly different cusum distribution will

Table 9   Percentiles of the parent distribution which correspond to given percentiles of the $(N-n+1)$st order statistic for sample size N

N	n	0·0001	0·001	0·01	0·50	0·99	0·999	0·9999
100	10	0·760	0·788	0·821	0·904	0·958	0·969	0·977
100	20	0·634	0·665	0·702	0·804	0·884	0·905	0·921
200	10	0·874	0·890	0·908	0·952	0·979	0·985	0·989
200	20	0·806	0·824	0·846	0·902	0·943	0·954	0·962
500	20	0·920	0·928	0·937	0·961	0·978	0·982	0·985
500	50	0·844	0·854	0·867	0·901	0·929	0·937	0·943

Note: the values in this table are only valid if the observations are taken independently from the parent distribution.

considerably raise the probability of the nth order statistic falling out-side the interval (l, u).

2.7   Cusum response to slow drifts

In particular cases a linear trend in the mean bias may occur. For example, one may have a forecast based on the assumption that average demand is a constant, while it actually has a linear trend. The situation also arises when one tries to forecast a quadratic trend with single exponential smoothing.

With a linear trend the bias will approach infinity. In the limit, therefore, the average sample number will be 1, the operating character-istic zero and hence the A.R.L. 1. In calculating the A.R.L. for a linear trend, one uses this limiting condition as a starting-point for the solution of the backward recurrence relations. At each step, the bias is reduced by the slope of the trend. The starting value of the bias has to be chosen high enough to suit the desired numerical accuracy of the final result.

These computations were carried out by the author for a particular example: Normal distribution of observations; cusum parameters $h = 2 \cdot 0$ and $k = 0 \cdot 4$; slope per period $0 \cdot 2$.

During each stage of development of the trend, the A.R.L. is short-er than that obtained when the same mean deviation from target exists permanently. In Table 10 the effect is shown for the above-mentioned

Table 10   A.R.L. of constant θ compared with A.R.L. for subsequent linear trend with slope 0·2 for cusum test ($h = 2 \cdot 0, k = 0 \cdot 4$)

θ	A.R.L.(θ)	A.R.L.(θ) when linear trend with slope 0·2
0·0	28·0	11·8
0·2	15·9	8·14
0·4	10·0	6·06
0·6	6·86	4·77
0·8	5·06	3·90
1·0	3·96	3·29
1·2	3·24	2·84
1·4	2·74	2·50
1·6	2·38	2·23

test. For each value of θ the A.R.L. for that θ is given, and also the A.R.L. if it is known that after this observation θ will increase by 0·2 per observation.

3 THEORY

3.1 Analytic solutions of the integral equations

Most of the remainder of this monograph is devoted to a brief discussion of the various ways in which more information about the behaviour of particular cusum tests can be obtained by solving the integral equations for $P(z)$, $N(z)$ and the A.R.L. $L(z)$, which were first mentioned in section 1.6.

These equations (27), (30) and (31) are inhomogeneous Fredholm equations of the second kind (see for example Lovitt (1924)). We shall in this section take the equation for the operating characteristic $P(z)$ as the example, all points being made being quite generally transferable to the equations for $N(z)$ and $L(z)$:

$$P(z) = F(-z) + \int_0^h f(x-z)\, P(x)\, dx \quad (0 \leqslant z \leqslant h). \tag{31}$$

We shall disregard the fact that the function F is related to f, as this is of no importance in the discussion of the integral equation and its solution. $f(x-z)$ is the kernel of the integral equation; it is both a probability kernel and a difference kernel. The fact that the equations have difference kernels, being only a function of $x-z$ rather than of the two variables x and z, has an interesting and useful consequence. An elegant way exists to obtain solutions for a group of integral equations with the same difference kernel but different "free terms F", without much additional work for each new equation. This is important in the present application, because P, N and L obey integral equations with the same difference kernel. Given the solution of the integral equation with the free term equal to the kernel, each similar equation with a different free term has a solution, which can be found by a single integration. We proceed to show this.

By repeated substitutions of the equation (31) into itself, it is possible to shift the position of the unknown function $P(x)$ in the right-hand side of the equation to a position where it has less weight. After the first substitution one obtains

$$P(z) = F(-z) + \int_0^h f(x-z)\left\{ F(-x) + \int_0^h f(x'-x)\, P(x')\, dx' \right\} dx \tag{41}$$

$$= F(-z) + \int_0^h f(x-z)\, F(-x)\, dx + \int_0^h P(x') \int_0^h f(x-z)\, f(x'-x)\, dx\, dx'.$$

54

Defining $\int_0^\hbar f(x-z)\, f(x'-x)\, dx = K_2\,(x'-z)$ as the second iterated kernel, we can write this as

$$P(z) \;=\; F(-z) + \int_0^\hbar K_1(x-z)\, F(-x)\, dx + \int_0^\hbar K_2(x-z)\, P(x)\, dx, \qquad (42)$$

in which we have written the original kernel $f(x-z)$ as the first iterated kernel $K_1(x-z)$. The iterated kernels of a difference kernel are easily shown to be, again, difference kernels. See for this and other more rigorous discussions of these problems, Courant and Hilbert (1953), Lovitt (1924), Mikhlin (1957). By repeating this procedure and defining further iterated kernels,

$$K_m(x-z) \;=\; \int_0^\hbar K_1(x'-z)\, K_{m-1}(x-x')\, dx',$$

we obtain the solution of the integral equation by

$$P(z) \;=\; F(-z) + \sum_{m=1}^\infty \int_0^\hbar K_m(x-z)\, F(-x)\, dx. \qquad (43)$$

Approximations are obtained by taking partial sums. Provided that the approximations converge uniformly in z, it is possible to interchange the order of summation and integration in (43), giving (the uniform convergence holds; see, for a proof, the references cited above):

$$P(z) \;=\; F(-z) + \int_0^\hbar F(-x) \sum_{m=1}^\infty K_m(x-z)\, dx. \qquad (44)$$

The function $G(x-z) = \sum_{m=1}^\infty K_m(x-z)$ is called the *resolvent* of the integral equation. It depends only on the kernel and range of integration, and not on the free term of the equation. Once the resolvent is known, it is possible to solve every equation with that particular kernel and range of integration, in which the only extra work involved is the one integration in (44).

In the case considered here, that of a difference kernel, the resolvent is the solution of the same type of integral equation, in which for the free term the kernel has been taken:

$$G(x) \;=\; K_1(x) + \int_0^\hbar K_1(s-x)\, G(s)\, ds$$

$$\;=\; K_1(x) + \int_0^\hbar K_1(s) \sum_{m=1}^\infty K_m(s-x)\, ds \;=\; K_1(x) + \sum_{m=2}^\infty K_m(x) \;=\; \sum_{m=1}^\infty K_m(x).$$

$$(45)$$

Hence, to solve (27), (30) and (31), one can first solve

$$G(z) = f(z) + \int_0^h f(x-z)\, G(x)\, dx$$

and then find the solutions of the equations by substituting the resolvent into expressions like (44). This may be important in numerical solutions, such as those discussed in section 3.7, in order to cut down on computing time.

Analytic solutions to these Fredholm equations can only be found for certain relatively simple kernels. One of the types of kernels that admits analytic solutions is the Gamma density, being a polynomial multiplied by an exponential. Actually, all sums of polynomials multiplied by exponentials are so-called degenerate kernels, which can be written as

$$K(x,z) = \sum_{i=1}^{n} A_i(x)\, B_i(z). \tag{46}$$

The kernels for Gamma distributions become

$$K(x-z) = f(x-z) = \frac{(x-z+p-m)^{p-1}}{(p-1)!} \exp(-x+z-p+m), \tag{47}$$

in which p is the Gamma shape parameter, m the mean of the Gamma distribution, and $\sqrt{p}$ its standard deviation. The formula (47) for the kernel is valid for $x-z \geqslant m-p$ and the kernel is zero elsewhere. After splitting (47) into the functions $A_i(x)$ and $B_i(z)$ of (46), the integral equation (31) is rewritten:

$$P(z) = F(-z) + \sum_{i=1}^{p} B_i(z) \int_0^h A_i(x)\, P(x)\, dx$$

$$= F(-z) + \sum_{i=1}^{p} B_i(z)\, C_i, \tag{48}$$

in which the C_i are constants, as yet unknown, defined by

$$C_i = \int_0^h A_i(x)\, P(x)\, dx \quad (i = 1, 2, \ldots, p). \tag{49}$$

The values of the constants C_i are found by substituting (48) into (49):

$$C_i - \sum_{j=1}^{p} C_j \int_0^h A_i(x)\, B_j(x)\, dx = \int_0^h F(-x)\, A_i(x)\, dx \quad (i = 1, 2, \ldots, p). \tag{50}$$

All integrals in (50) are known, and after computing these the C_i are found as the solution of a set of linear equations with p unknowns. These solutions substituted into (48) give $P(z)$.

This method of solution as applied to Gamma kernels can only be used when $m-p \leqslant -h$, because otherwise the C_i become functions of z, which themselves obey integral equations. Another restrictive condition is that p must be an integer.

The solutions for example 2 in section 1.6 can readily be found using this method with $p=1$, $m=0$ and $h=1$. In that special case there is only one constant C_i.

Although it is thus possible in principle to evaluate the solutions of the integral equations for certain Gamma kernels, one need not necessarily take this road. Even this "analytic" solution requires that $p(p+1)$ integrals have to be evaluated and the solution of p linear equations with p unknowns has to be found, which may be prohibitive for large or moderately large p, so one may still have to resort to numerical techniques to accomplish this. In that case it might well be more efficient and direct to use the numerical methods of section 3.7.

3.2   Approximate solutions I: the Neumann series

The next four sections briefly indicate various approximate ways of dealing with the integral equations, which may yield numerical results for the A.R.L. and other quantities that are of interest. The common title "Approximate solutions" is no reflection on the accuracy of these methods in general as compared with the analytic solutions mentioned in the previous section. Indeed, as was indicated at the end of the previous section, it is not easy to draw a line between analytic and numerical solutions. In so far as the solutions of 3.1 differ from the others, they do so in that they lead to a closed expression for the solutions of the integral equations (writing the C_i as the quotients of Cramer determinants).

We have seen in the foregoing section, in (43), that the solution of integral equation (31) can be written as an infinite sum of integrals, using the iterated kernels $K_m(x-z)$. Rewriting (43) slightly shows how approximations to the solution may be obtained:

$$P(z) = F(-z) + \sum_{m=1}^{n} \int_0^h K_m(x-z)\,F(-x)\,dx + \int_0^h K_{n+1}(x-z)\,P(x)\,dx$$

$$= P_n(z) + R_n(z), \tag{51}$$

in which

$$P_n(z) = F(-z) + \sum_{m=1}^{n} \int_0^h K_m(x-z)\,F(-x)\,dx \tag{52}$$

and

$$R_n(z) = \int_0^h K_{n+1}(x-z)\,P(x)\,dx. \tag{53}$$

The $P_1(z)$, $P_2(z)$, ... form successive terms of a series, the

Neumann series, which converges uniformly to $P(z)$. As F and f, and hence K_m, are all non-negative functions throughout their range, the series is in this case also a monotonically non-decreasing function of n for each z $(0 \leqslant z \leqslant h)$. As the limit function of the Neumann series is continuous in z for continuous kernels, this monotonicity proves the uniform convergence.

Lower limits on the solution can thus be given to an arbitrary degree of accuracy by evaluating $P_n(z)$ for increasing n. A lower limit on the error of the nth approximation is given by

$$\int_0^h K_{n+1}(x-z)\, P_n(x)\, dx. \tag{54}$$

If

$$\int_{-h}^h f(x)\, dx = A < 1, \tag{55}$$

which will always hold for random variables with a semi-infinite or infinite domain, an upper limit to the error of approximation can also be given by finding an upper limit for

$$R_n(z) = \sum_{m=n+1}^{\infty} \int_0^h K_m(x-z)\, F(-x)\, dx = \int_0^h F(-x) \sum_{m=n+1}^{\infty} K_m(x-z)\, dx. \tag{56}$$

If (55) holds, then

$$K_m(x-z) \leqslant A^{m-n} \left| \int_0^h K_n(x-x')\, dx' \right| \quad (m = n+1,\, n+2,\, \dots). \tag{57}$$

Hence (56) can be limited from above:

$$R_n(z) \leqslant \frac{A}{1-A} \int_0^h F(-x) \left| \int_0^h K_n(x-x')\, dx' \right| dx. \tag{58}$$

Whether or not this is a useful upper bound depends on the value of A.

An interesting example is again provided by example 2 of section 1.6. We work this out for $N(z)$, for which the integral equation was

$$N(z) = 1 + \int_0^1 e^{-1} \exp(z-x)\, N(x)\, dx = 1 + e^{-1} \exp(z).$$

The iterated kernels are $K_m(x-z) = e^{-m} \exp(z-x)$. Hence successive terms of the Neumann series are

$$N_m(z) = 1 + \sum_{i=1}^{m} e^{-i}(1 - e^{-1})\, \exp(z)$$

$$= 1 + e^{-1} \exp(z) - e^{-m-1} \exp(z) = N(z) - e^{-m-1} \exp(z).$$

Using estimate (54) of the error and adding this to the approximation $N_m(z)$, the error is reduced to $- e^{-2m-2} \exp(z)$, which speeds up convergence considerably.

58

3.3 Approximate solutions II: approximating the equation

The limited use that can be made of analytic tools in solving the cusum integral equations is mainly due to the finite interval of integration $(0, h)$. An infinite or even semi-infinite interval of integration would have made much more general solutions available to us for quite general kernels (without limitation to degenerate kernels, etc.).

A method of finding approximate solutions to the integral equations is therefore to make certain assumptions about the unknown function outside its interval of definition $(0, h)$ and to make the interval of integration infinite.

This method was used by K.W. Kemp (1961) and seemed to give close approximations for sufficiently large h. We follow through Kemp's argument first for the operating characteristic and then for the average sample number. In

$$P(z) = \int_{-\infty}^{0} f(x-z)\,dx + \int_{0}^{h} f(x-z)\,P(x)\,dx \qquad (31)$$

the first term of the right-hand side can be replaced by

$$\int_{-\infty}^{0} f(x-z)\,P(x)\,dx,$$

if $P(z) \approx 1{\cdot}0$ for $z \leqslant 0$, which is certainly the case if the mean of the distribution of accumulated values is small or negative. We further assume that the integral $\int_{h}^{\infty} f(x-z)\,P(x)\,dx$ is negligible compared to the value of $\int_{0}^{h} f(x-z)\,P(x)\,dx$, because

   (1) when μ is large, $P(z) \approx 0$ for $z \geqslant h$

   (2) when μ is small, $\int_{h}^{\infty} f(x-z)\,dx \approx 0$ for $z \geqslant h$.

We therefore approximate equation (31) by

$$P'(z) = \int_{-\infty}^{\infty} f(x-z)\,P'(x)\,dx. \qquad (59)$$

This is now a homogeneous integral equation with an infinite interval of integration. This equation always has the trivial solution $P'(z) \equiv 0$. If the homogeneous equation has several solutions other than the trivial solution, all linear combinations of these solutions are also solutions of the equation. It can immediately be seen for (59) that $P'(z) \equiv 1$ is a non-trivial solution, because $\int_{-\infty}^{\infty} f(x-z)\,dx = 1$ for all z; so $P'(z) \equiv C$ is also a solution, in which C is any constant.

Another solution of this equation is $P'(z) = \exp(az)$, if a is such that

$$\int_{-\infty}^{\infty} f(x)\exp(ax)\,dx = 1, \tag{60}$$

as can readily be verified by substitution into the integral equation. See Titchmarsh (1937) for a proof that all solutions of (59) are of this type. The left-hand side of (60) is the moment-generating function of $f(x)$. We are therefore interested in non-zero roots of (60); the zero root gives $P'(z) \equiv C$. Having obtained such a root, we then write the solution of the homogeneous approximate equation (59) as

$$P'(z) = C + D\exp(az). \tag{61}$$

As the solution of the homogeneous equation allows two arbitrary constants C and D, we try to choose these such that the solution fits the original integral equation as closely as possible. Substituting first $z = 0$ and then $z = h$ into (61), we obtain

$$P'(0) = C + D \tag{62a}$$

$$P'(h) = C + D\exp(ah). \tag{62b}$$

Hence

$$C = \frac{P'(h) - P'(0)\exp(ah)}{1 - \exp(ah)} \tag{63a}$$

$$D = \frac{P'(0) - P'(h)}{1 - \exp(ah)}. \tag{63b}$$

Substituting these into (61), we find

$$P'(z) = P'(0) - \left\{P'(0) - P'(h)\right\}\frac{1 - \exp(az)}{1 - \exp(ah)}. \tag{64}$$

Substituting this into the right-hand side of the original equation (31) and using first $z = 0$ and then $z = h$, we obtain two linear equations in $P(0)$ and $P(h)$. The solutions from these into (64) give the approximation for $P(z)$.

Working this out for the special case of the Normal distribution with mean μ and standard deviation equal to unity, we find that (60) is satisfied for $a = -2\mu$. (64) becomes

$$P'(z) = P'(0) - \left\{P'(0) - P'(h)\right\}\frac{1 - \exp(-2\mu z)}{1 - \exp(-2\mu h)}. \tag{64a}$$

Substituting this into (31) for $z = 0$ and $z = h$ gives the following two equations for $P(0)$ and $P(h)$:

$$K_1 P(0) + K_2 P(h) = F(-\mu) \tag{65a}$$

$$\exp(-2\mu h) K_2 P(0) + K_1 P(h) = F(-h-\mu) \tag{65b}$$

with

$$K_1 = 1 - F(h-\mu) + F(-\mu) - K_2 \tag{66a}$$

$$K_2 = \frac{1 + F(h+\mu) - F(h-\mu) - 2F(\mu)}{1 - \exp(-2\mu h)}. \tag{66b}$$

For $\mu = 0$ some difficulties occur, which have to be dealt with by a limiting process:

$$P(0) = \frac{\tfrac{1}{2}h\sqrt{2\pi} + \left\{1 - \exp(-\tfrac{1}{2}h^2)\right\}}{h\sqrt{2\pi}\left[\tfrac{3}{2} - F(h)\right] + 2\left\{1 - \exp(-\tfrac{1}{2}h^2)\right\}} \tag{67}$$

and

$$P(z) = P(0) + \left\{1 - 2P(0)\right\}z/h. \tag{68}$$

These approximations have been shown to give fairly accurate results for $h = 10$ by Kemp. They are not considered suitable, however, for use with $h \leqslant 3$, because there is no check possible as to their accuracy. It is remarkable, however, how close some values are even for small h. For example, for $\mu = 0$, $h = 2$ we find $P(0) \approx 0\cdot778$ for a true value of $0\cdot776$. For $\mu = 0\cdot4$, $h = 3$ we obtain $P(0) \approx 0\cdot51$ as against a true value of $0\cdot535$.

For the average sample number $N(z)$, Kemp finds a very similar approximation by using

$$N'(z) = 1 + \int_{-\infty}^{\infty} f(x-z)N'(x)\,dx \tag{69}$$

on the grounds that $N(z) \approx 1$ for $z \leqslant 0$ and $F(-z)$ and $1 - F(h-z)$ are small enough to be neglected.

This yields the solution

$$N'(z) = \frac{\{C + D \exp(az)\} - z}{\mu} \tag{70}$$

in which a has the same meaning as before. A similar process of substitutions, as shown for $P(z)$, gives the approximate solution.

3.4 Approximate solutions III: Wiener-Hopf techniques

The Fourier transform analysis can be used to solve integral equations with an infinite domain of integration — see Titchmarsh (1937). The Wiener-Hopf technique extends the use of Fourier transform methods to the solution of integral equations with a semi-infinite domain of integration. See Wiener and Hopf (1931), Noble (1958), Titchmarsh (1937), Feller (1966), Kemperman (1963), Spitzer (1957).

Our interest, however, is in the finite domain of integration $(0, h)$. Here a Wiener-Hopf type approach is possible with some added complication. The problem becomes the so-called "three part problem" (Noble (1958)). Latter (1958) has given an extension of the Wiener-Hopf method that makes it possible to obtain approximations to the solution of integral equations (30) and (31). The accuracy of the approximations improves exponentially with increasing order of approximation.

Both Feller (1966) and Keilson (1965) remark that Wiener-Hopf methods are a rather too general analytical tool to do full justice to the probabilistic nature of the problem. Little insight is offered into the statistical nature of the process, and the use of probability kernels in the integral equations makes a simplification of the general methods possible — see e.g. Spitzer (1957).

The method given by Latter (1958) only applies to a symmetric kernel, i.e. when $f(x-z) = f(z-x)$ in (30) and (31). This is only the case in a limited number of instances, e.g. for the normal distribution with mean zero.

We think that these particular techniques have not been taken far enough at present to yield practical solutions for a large class of interesting cases of equations (30) and (31). We therefore limit ourselves to the formal argument, which leads up to a Wiener-Hopf type equation.

The integral equation

$$P(z) \; = \; F(-z) + \int_0^h f(x-z) \, P(x) \, dx, \tag{31}$$

is only valid for z in the interval $(0, h)$. We now extend the region of validity to the infinite interval $(-\infty, \infty)$; in fact, we can define $P(z)$ arbitrarily outside $(0, h)$ and so choose the definition that gives the best chance of being useful to the solution.

First we rewrite (31) as

$$\int_0^h K(z-x) \, P(x) \, dx \; = \; F(-z), \quad (0 \leqslant z \leqslant h) \tag{71}$$

in which

$$K(z-x) \; = \; f(x-z) - \delta(x-z) \tag{72}$$

and

$$-F(-z) \; = \; \int_{-\infty}^0 f(x-z) \, dx \tag{73}$$

(δ is the Dirac delta function).

We further define the functions $D(z)$ and $E(z)$ to extend $P(z)$ below and above the interval $(0, h)$:

$$D(z) \; = \; \int_0^h K(z-x) \, P(x) \, dx \quad (h < z < \infty), \tag{74}$$

62

and

$$E(z) = \int_0^h K(z-x)P(x)\,dx \qquad (-\infty < z < 0). \tag{75}$$

Now multiply both sides of (71) by $\exp(i\alpha z)/\sqrt{2\pi}$ and integrate over z from $-\infty$ to ∞.

$$\frac{1}{\sqrt{2\pi}} \int_{-\infty}^{\infty} \exp(i\alpha z) \int_0^h K(z-x)P(x)\,dx\,dz =$$

$$= \frac{1}{\sqrt{2\pi}} \int_0^h P(x) \int_{-\infty}^{\infty} K(z-x)\exp(i\alpha z)\,dz\,dx =$$

$$= \Psi_-(\alpha) + C(\alpha) + \Phi_+(\alpha), \tag{76}$$

where

$$\Psi_-(\alpha) = \frac{1}{\sqrt{2\pi}} \int_{-\infty}^{0} E(z)\exp(i\alpha z)\,dz \qquad \text{(unknown)} \tag{77}$$

$$C(\alpha) = \frac{1}{\sqrt{2\pi}} \int_0^h F(-z)\exp(i\alpha z)\,dz \qquad \text{(known)} \tag{78}$$

$$\Phi_+(\alpha) = \frac{1}{\sqrt{2\pi}} \int_h^{\infty} D(z)\exp(i\alpha z)\,dz \qquad \text{(unknown)}. \tag{79}$$

The left-hand member of (76) can be written ($y = z - x$) as

$$\frac{1}{\sqrt{2\pi}} \int_0^h P(x)\exp(i\alpha x)\,dx \cdot \int_{-\infty}^{\infty} K(y)\exp(i\alpha y)\,dy = -\Phi_1(\alpha)H(\alpha) \tag{80}$$

where

$$\Phi_1(\alpha) = \frac{1}{\sqrt{2\pi}} \int_0^h P(x)\exp(i\alpha x)\,dx \tag{81}$$

and

$$-H(\alpha) = \int_{-\infty}^{\infty} K(y)\exp(i\alpha y)\,dy. \tag{82}$$

The Wiener-Hopf type equation we end up with is then

$$\Phi_+(\alpha) + \Psi_-(\alpha) + H(\alpha)\Phi_1(\alpha) + C(\alpha) = 0. \tag{83}$$

This equation is valid in the strip $\tau_- < \text{Im}(\alpha) < \tau_+$, if $\Phi_+(\alpha)$ is analytic in the half-plane $\text{Im}(\alpha) > \tau_-$ and $\Psi_-(\alpha)$ is analytic in the half-plane $\text{Im}(\alpha) < \tau_+$. There are three unknown functions Φ_+, Ψ_- and Φ_1 and two known functions $H(\alpha)$ and $C(\alpha)$. Further information that can be exploited is $\lim_{z \to -\infty} P(z) = 1$, $\lim_{z \to \infty} P(z) = 0$ and $P(z)$ is a non-increasing function of z.

3.5 Approximate solutions IV: renewal-theoretic considerations

An often used approximate formula for the average run length $L(0)$ of a cusum test on normal observations is

$$L(0) \approx R + h/(\theta - k), \tag{84}$$

in which $\theta - k$ is the mean of the distribution of the values to be accumulated in the cusum and R is a constant, for which slightly differing values are given by different sources. R is found empirically to lie between $0 \cdot 5$ and 1 for $\theta - k$ positive and not too small.

Renewal theory supports this approximation. Renewal theory, the study of certain aspects of random walks with non-negative increments, is described in Cox (1962) and Feller (1966). An extension of one result to the general case of increments which can also be negative is given by Feller and Orey (1960).

In renewal theory one may study the behaviour of electric light-bulbs, for example. One light-bulb is switched on; when it fails, it is immediately replaced by a new one: this process is repeated at the second failure and so on, *ad infinitum*. One is then interested in making probability statements about the number of renewals up to a certain time t, and also about the residual life-time of the bulb burning at time t. Other names used for residual life-time are "forward recurrence-time" and "excess"; we also speak of the time of the first renewal past t as the "hitting-point" of the random walk for the interval (t, ∞); see Feller (1966).

There is an analogy between the cumulative process in cusum tests and that in renewal theory, which is, however, obscured by the fact that the increments of the random walk are only seldom strictly non-negative for the cusum test. Exceptions to this rule exist, for example when $\theta - k$ is greater than $\sqrt{p}$ in a Gamma distribution of shape parameter p.

We may expect, however, that values for the A.R.L. computed by renewal theoretic methods will be fairly accurate, when one may neglect the negative part of the probability distribution of the values to be accumulated.

The analogy is fairly direct, replacing successive life-times of light-bulbs by successive values to be accumulated in the cusum and taking h for t. The A.R.L. is then represented by the mean number of renewals in the interval $(0, h)$ plus one, because one value of the cusum outside the interval $(0, h)$ is needed to terminate the cusum test.

To know the mean of the residual life-time distribution or excess distribution is to know the average run length, for we have, from continuity considerations:

$$\text{A.R.L.} = \frac{h + \text{expected excess}}{\theta - k} = \frac{h}{\theta - k} + \frac{\text{expected excess}}{\theta - k}. \quad (85)$$

This formula is of the form (84) and one may expect to find the proper value of R to use by considering the expected excess. Renewal theory gives an asymptotic formula for the expected excess; for large h the expected excess approaches $(\mu^2 + \sigma^2)/2\mu$, in which μ is the mean of the life-time distribution and σ its standard deviation; for the cusum test the standard deviation is 1, as we are using standardized deviations from target, and the asymptotic expected excess becomes

$$\frac{(\theta - k)^2 + 1}{2(\theta - k)} = \frac{\theta - k}{2} + \frac{1}{2(\theta - k)}. \quad (86)$$

From (85) and (86) one would therefore expect the value $R = 0{\cdot}5$ to be appropriate for large h, large $\theta - k$, and a higher value to be appropriate for large h, smaller $\theta - k$. In Table 11 we show the values of R

Table 11   Values of $L(0) - \dfrac{h}{\theta - k} = R$ for the Normal distribution

$\theta - k$	$h = 0$	$h = 2$	$h = 4$	$h = 6$	$h = 8$	$h = 10$	$0{\cdot}5 + 0{\cdot}5/(\theta - k)^2$
0·4	1·53	0·06	−0·12	−0·14	−0·12	−0·10	3·62
0·8	1·27	0·74	0·74	0·74	0·75	0·73	1·28
1·2	1·13	0·71	0·72	0·72	0·72	0·72	0·85
1·6	1·06	0·64	0·66	0·66	0·66	0·66	0·70
2·0	1·02	0·58	0·62	0·62	0·62	0·62	0·62
3·0	1·00	0·49	0·59	0·54	0·56	0·56	0·56
4·0	1·00	0·52	0·50	0·56	0·51	0·55	0·53

as a function of h and $\theta - k$, which would make the approximation exact locally for the normal distribution. It can be seen that $0{\cdot}5 + 0{\cdot}5/(\theta - k)^2$ is a fairly good approximation for these values, even for small h.

This sufficiently justifies the use of (84) locally. In interpolating in the table of A.R.L., it is one of the most useful aids for $\theta - k > 0{\cdot}5$. Using the estimated values for R for the surrounding values of h and $\theta - k$, one can interpolate for R linearly and so calculate the A.R.L.

It is possible to use further results as obtained by Feller and Orey (1960) to estimate the difference between the A.R.L. for a Normal distribution and, say, a Gamma distribution by taking the differences of the generalized renewal densities integrated over $(0, h)$. This extends the useful range of Tables 5 outside the Normal distribution for these particular values of the parameters.

The theorem given by Feller and Orey (1960) is valid for general random walks. It gives an expression representing the expected number of points of the random walk in a certain interval. The A.R.L. of a cusum test would be given exactly by the expression given in the theorem, if no returns to the origin occurred. It is in these returns to the origin that the cusum test quantity differs from ordirary random walks. In the cusum test each excursion into the negative half-axis causes a jump to the origin. This, however, leads us to believe intuitively that the quantity given by the theorem will be a good approximation for sufficiently high $\theta - k$, where the slight inaccuracy caused by occasional returns to the origin may be neglected.

The author has tried this method of approximating the A.R.L. for the same parameter values as those of Table 11 and the results are very promising indeed. Not only are very accurate approximations obtained for any h (not only valid for high h), but they are also useful for smaller values of $\theta - k$. In addition to this they provide a relatively easy way of obtaining estimates of the difference between the A.R.L. for some distribution other than the Normal, and the Normal distribution itself.

Let us define $F^{k^*}(x)$ as the distribution function of the kth self-convolution of the values to be accumulated; $F^{0^*}(x) \equiv 1$ and $F^{1^*}(x) = F(x)$.

Without going into the restrictive conditions of the theorem or showing the general form of the theorem, which may be found in Feller (1966), the approximation to the A.R.L. is

$$\hat{L} = 1 + \sum_{k=0}^{\infty} F^{k^*}(h) - \sum_{k=0}^{\infty} F^{k^*}(0). \tag{87}$$

It was found that the approximation was very good. Rather than give a whole table here, we restrict ourselves to mentioning the per cent error of the approximation for some entries. These can easily be checked by the reader, as only tables of the integral of a Normal probability density and some additions are required. The number of additions needed to obtain the infinite sums of (87) to a sufficient degree of approximation depends mainly on $\theta - k$. If for $\theta - k = 0.5$ a hundred values are needed, one need take only ten for $\theta - k = 2$. The author has found similar accuracies to obtain for the Gamma distributions used in Table 7 in section 2.4.

For the Normal distribution the per cent error of the "renewal approximation" is less than one per cent of the true A.R.L. for $h = 2, \theta - k \geqslant 1.8$; $h = 4, \theta - k \geqslant 1.4$; $h = 6, \theta - k \geqslant 0.8$; $h = 8, \theta - k \geqslant 0.6$; $h = 10, \theta - k \geqslant 0.6$.

For $h = 2$, for $\theta - k$ even as far down as 0·6, the error is less than 3 per cent of the A.R.L.

3.6   Approximate solutions V: Monte Carlo methods

If a digital computer is available, there is a very easy way to obtain the A.R.L. of cusum tests, namely by simulating a sufficient number of runs. Observations having the required distribution can be generated and the cusum test (numerical version) applied to it. As soon as a signal occurs, the run length is counted into a frequency table if one is also interested in the run length distribution, and the next run is made after resetting the cusums.

Advantages of this approach are: (1) the computer program is easy to write — one or two days' work may produce a running program; (2) standard routines for random number generation are usually available; (3) the run length distribution can be obtained as a cheap by-product of the simulation; (4) the accuracy of the resulting estimate of the A.R.L. can be determined reasonably well, although it is in probability terms.

There is one serious disadvantage of Monte Carlo methods if one desires a fairly high accuracy, i.e. the enormous amount of computer time that is needed. Especially for cusum parameters which cause the A.R.L. to be high, the number of computational steps required is very large. We illustrate this with an example. An approximate random number generator for Normal observations is used, which uses 20 multiplications per random number generated. The A.R.L. to be estimated is about 500 and the standard deviation of the run-length distribution is also 500. With a probability of 99 per cent the result should be within 0·5 per cent of the true value. This means that the standard error of the estimate should not be more than about 0·2 per cent of the estimate itself. The standard error of the estimate of a mean of a distribution is the standard deviation of that distribution divided by the square root of the number of observations in the sample. As the standard deviation equals the mean here, the standard error of the estimate of the A.R.L. is the estimate itself divided by the square root of the number of runs. Then for the desired accuracy about 250,000 runs must be made of 10,000 multiplications each, totalling 2,500,000,000 multiplications for the one A.R.L. Even on a very high-speed computer this becomes expensive.

For the Normal distribution, typical times for the Control Data 3600 computer were (·12 times A.R.L.) minutes per 10,000 runs. The above example would have taken about 25 hours on that very fast machine. Very few research institutions would be prepared to compile tables like those in section 2.2 of this monograph on that basis.

We summarize by stating that we consider Monte Carlo methods as "brute force" methods, which may be used for first approximate results, to be obtained quickly. But for the compilation of accurate tables the method is far inferior to the numerical method of the next section and, in the cases where this is applicable, even to the approximate methods of preceding sections. Another good field for application of Monte Carlo methods is where the derivation of simple theoretical expressions such as the integral equations is impossible (see section 2.5 for an example of this).

Naturally, the simulation method is not restricted to rectangular or normal random variables, but elegant ways exist to obtain random variables which are exactly Gamma-distributed: for some examples see Brown (1959), Jöhnk (1964), Tocher (1963).

3.7   Accurate numerical solutions using a digital computer

In this section, the method is given whereby it is possible to obtain accurate solutions of the integral equations (27), (30) and (31) at a reasonable cost of computation. The programming — if it is to be done on a digital computer — is rather more complex than that of the simulation described in the preceding section, but this is acceptable if a large number of A.R.L. values are required. This section is also the last of those discussing means of solving the integral equations. The discussion is far from complete, and in each section only an idea is given of the general approach.

The method underlying the numerical solution of the integral equations is that of *successive approximations*. The most interesting feature of this procedure is that one can start with almost any approximation and then improve this. If, therefore, some information on the solution already exists, this may be used to speed up the computations. On the other hand, if no such information is available, one can still arrive at the solution from as simple a starting solution as one can imagine (e.g. $N(z) \equiv 1$ or $P(z) \equiv 0$).

Taking again the equation for the operating characteristic (31) as the example, successive approximations to $P(z)$ are obtained according to

$$P_{n+1}(z) = F(-z) + \int_0^h f(x-z)P_n(x)\,dx \quad (0 \leqslant z \leqslant h), \ (n = 0, 1, \ldots).$$

$$(88)$$

The starting approximation is $P_0(z)$. The method thus simply consists of substituting an approximation into the right-hand side of the equation, so obtaining a better approximation. This is not necessarily a numerical process; it is possible with a not-too-complicated kernel to start with some function $P_0(z)$ and then analytically carry out some iterations as above.

Usually, however, the successive approximations are found by a numerical process. Then the integral in (88) also has to be replaced by an approximating finite sum. Various standard numerical methods are available here, such as Simpson integration, or more efficiently, Gauss integration. This in fact is equivalent to approximating the integral equations by a set of linear equations in m unknowns, in which m is the number of subdivisions in the integration formula ($m-1$ intervals). These linear equations are solved iteratively. Convergence is linear, i.e. the error is reduced by a fixed factor per iteration as one gets near the solution.

Simple stopping rules for terminating the iterations can be used, such as the following. Estimate the convergence factor λ at each step after the change at each step has become less than a predetermined value ϵ_1. The estimate of the distance from the correct solution is then $\lambda/(1-\lambda)$ times the latest change in solution. If this quantity becomes less than a predetermined value ϵ_2, computations are stopped for the number of subdivisions used. ϵ_2 is an estimate of the maximum error tolerable and can be chosen to suit the desired accuracy. The measure of change between two solutions can be several things, e.g. the change at $z = 0$, the maximum change for all z, or the sums of absolute values of the changes for all z used.

One can estimate the error in the final result by estimating the error in the finite sum approximation to the integral using the error formula for the integration formula used, but it is sometimes simpler to double the number of subdivisions and repeat the iterations until convergence occurs and then compare the result with that obtained for the original number of subdivisions. This will usually give a good indication of the residual error present. The number of iterations needed in a second stage is usually small, because interpolated values from the first solution provide good starting solutions. The computing time for the second stage per iteration will be about four times as long as that needed per iteration for the first stage.

One can of course go to great lengths in trying to optimize the procedure for the accuracy needed by having a substantially higher ϵ_1 and ϵ_2 for the first stage than those for the second stage. If the difference from the second stage is too great, then too many iterations will be needed in the second stage; if the difference is too small, the extra iterations in the first stage do not contribute to the speed of the second stage, because the extra accuracy of the solution to the first set of linear equations is made useless by the difference in the solutions to the first set of equations and the second set. The relationship between the ϵ's would then require a fairly good knowledge about the expected difference between the two final solutions.

More than two stages are possible of course, using the techniques described in the previous paragraph. One could go on until the difference between two successive solutions (of successive stages) is less than a predetermined amount, in much the same way in which the criterion is handled within a stage.

Most of the calculations for the tables of the A.R.L. for a Normal distribution given in this monograph have been computed using up to 129 subdivisions and utilizing either 48-bit or double precision 96-bit floating point arithmetic. Some computations have been done with up to 1025 subdivisions to check the accuracy and determine the pattern of solutions of successive stages.

When doing these numerical evaluations of the A.R.L., numerical values for the Normal probability density and its integral are needed. The density can be evaluated exactly, but an approximation is needed for the integral if one does not want to integrate every time a value is needed, which would be very time-consuming indeed. The author has used one of the approximations to the Normal probability integral given by Hastings (1955). The error of this approximation is less than plus or minus $0 \cdot 00000015$ in the whole positive range of x (the values for negative x can be derived from those for corresponding positive values). The approximation is (for a standardized normal distribution)

$$F(x) = 1 - 0 \cdot 5 / \left\{ 1 + a_1 x + a_2 x^2 + a_3 x^3 + a_4 x^4 + a_5 x^5 + a_6 x^6 \right\}^{16}, \tag{89}$$

in which
$$
\begin{aligned}
a_1 &= 0 \cdot 0498673470, & a_4 &= 0 \cdot 0000380036, \\
a_2 &= 0 \cdot 0211410062, & a_5 &= 0 \cdot 0000488906, \\
a_3 &= 0 \cdot 0032776263, & a_6 &= 0 \cdot 0000053830.
\end{aligned}
$$

In programming this approximation for a digital computer, being in the inner loop of the computations if we are solving the equation for $P(z)$, one must try to make the routine for this as fast as possible by using Horner's method to evaluate the sixth-degree polynomial and obtain the 16th power by squaring repeatedly. It is even better to avoid having these evaluations in the inner loop of the computations by solving only the resolvent equation (see section 3.1), which only uses the probability density. The solutions to the equations for $P(z)$ and $N(z)$ are then obtained by one numerical integration (44, section 3.1); for $P(z)$ this involves the probability integral, but it now occurs only once instead of in each iteration.

For Gamma distributions, the evaluation of both density and integral can be exact, although the computing time goes up steeply for increasing shape parameter (this has to be an integer for easy evaluation).

The problem of the accuracy of the A.R.L. becomes serious for very high A.R.L. This is usually caused by $P(0)$ being very close to 1. Very small errors in the computed value for $P(0)$ can then cause excessive errors in the A.R.L. The error ΔL in $L(0)$ as a function of ΔP, the error in $P(0)$, and of ΔN, the error is $N(0)$, is, for sufficiently small ΔP and ΔN, approximately

$$\Delta L \approx \frac{N(0)}{\{1 - P(0)\}^2} \Delta P + \frac{1}{1 - P(0)} \Delta N. \tag{90}$$

As the errors in $N(0)$ and $P(0)$ will usually be of the same order of magnitude, using the method of solution described here, the first term on the right-hand side of (90) usually dominates the second. This is especially the case when $P(0)$ is close to 1. It is now clear why it is sometimes difficult to say whether an A.R.L. is 10^8 or 10^9, because ΔP would have to be less than about 10^{-14} to discriminate between the two. Fortunately one is normally not too worried about the exact value of the A.R.L. at such heights, as long as one can make a statement such as: the A.R.L. is at least 10^8.

Much less difficulty is encountered in the calculation of lower A.R.L., where convergence is quicker, because less accuracy is needed for $P(0)$ for a given desired accuracy of the A.R.L. (in the author's program, the accuracy of $P(0)$ is automatically adjusted to give a specified relative accuracy of the resulting A.R.L.), and also good and simple approximations are often available to provide good starting solutions (see section 3.5). To start up the solution for high A.R.L., the best values to use are those given by the approximations by Kemp (1961) – see section 3.3.

The recurrence equation (88) also has a meaning for non-stationary processes such as the step-response of exponential smoothing (section 2.5). If the function $P_n(z)$ is the correct operating characteristic at time t, then $P_{n+1}(z)$ is the correct operating characteristic for the test at time $t-1$. (88) is therefore a backward recurrence relation valid for non-stationary processes, if n is replaced by $t-n$ and $n-1$ by $t-n-1$ and also f by f_{t-n-1}, the probability density at time $t-n-1$; similarly for F_{t-n-1}. This property has been used to calculate the A.R.L. for step-responses of exponential smoothing and for linear trends by first calculating a "steady state" solution for $\theta = 0$ and $\theta = \infty$, respectively, and then tracing backwards through time to zero time, using (88).

The results of a double precision calculation (96-bits floating point arithmetic) of five successive stages for the normal distribution and cusum parameters $h = 5$, $\theta - k = -2$ (not shown in our tables) were:

No. of subdivisions	No. of iterations	A.R.L.
40	11	too large
60	17	$10 \cdot 67 \times 10^6$
90	22	$8 \cdot 45 \times 10^6$
135	26	$8 \cdot 24 \times 10^6$
200	30	$8 \cdot 21 \times 10^6$

3.8 Run-length distributions

The distribution of the run length of a cusum test is even more difficult to obtain analytically or numerically than the average run length. Although we shall use the integral equation for the run-length distribution as given in Ewan and Kemp (1960), this is mainly of theoretical interest. For high A.R.L., $P(0)$ is usually high, which means that the high A.R.L. is caused by a high average number of repetitions of the test rather than by high average sample numbers. As the distribution of the number of repetitions is geometric, the distribution of the run length will also be very nearly geometric, except for very low run lengths. We then approximate the probability $p(n, 0)$, that a test starting at 0 will have a run length of n, by

$$p(n, 0) \; = \; 1/L(0) \cdot \exp\{ - (n-1)/L(0)\}. \tag{91}$$

For low A.R.L. probably the best way to obtain information about the run-length distribution is to use Monte Carlo methods, which take an amount of computation proportional to the A.R.L. (section 3.6).

If we define $p(n, z)$ as the probability that a run length of n will result in a test starting at z, the integral equation for $p(n, z)$ is (Ewan and Kemp (1960)):

$$p(n, z) \; = \; p(n-1, 0)\, F(-z) \; + \int_0^h p(n-1, x)\, f(x-z)\, dx \tag{92}$$

$$0 \leqslant z \leqslant h, \; n = 2, 3, \ldots$$

As

$$p(1, z) \; = \; 1 - F(h - z), \tag{93}$$

it is possible to solve (92) successively for $n = 2, 3, \ldots$ by analytic or numerical integration. For low A.R.L. it even provides an interesting alternative way to calculate the A.R.L., by solving (92) for successive n and terminating when there is no further significant contribution to the first moment of $p(n, 0)$.

Unfortunately, this method may also be advantageous only for relatively low A.R.L., for which case there are sufficient alternative good solutions. For large A.R.L., one has the serious disadvantage that a build-up of rounding errors occurs, which does not happen with the self-correcting convergent procedures of the preceding section.

It is possible to derive integral equations for the moments of the run-length distribution by differentiating the integral equation for the moment-generating function of the run-length distribution, which follows from (92). The first result is then the integral equation for $L(z)$, which was found in section 1.6 (27). The equation for the second zero moment of the run-length distribution is

$$\mu_2'(z) = 2L(z) - 1 + \mu_2'(0) F(-z) + \int_0^h \mu_2'(x) f(x - z) \, dx. \qquad (94)$$

Appendix: **CONDITION ON CUSUM TEST PARAMETERS FOR (21) TO BE TRUE**

Kemp (1961) proved that for a symmetrical two-sided cusum test the following statement is true:

"A signal occurring on the upper (lower) boundary of a cusum test will not interrupt a test, which might have ended on or beyond the lower (upper) boundary, because the test quantity of the test on negative (positive) deviations is zero at the time of the signal." (A1)

Kemp (1961) also stated that statement (A1) is true for asymmetrical tests with $d^+ = d^- = d$, if

$$\sin(\phi_1 + \phi_2) > d \sin(\phi_1 - \phi_2), \quad \text{for } \phi_1 > \phi_2.$$

We shall proceed to find a necessary and sufficient condition for the parameters of a two-sided test, using essentially the method of proof of Kemp (1961) adapted to our purpose. One should note that of course h^+, h^-, k^+, k^- are all non-negative, which is implicitly assumed in the proof.

Let us consider a situation in which, immediately after taking the rth observation, both test quantities are within their limits, whereas after the $(r+1)$th observation the positive test quantity exceeds h^+, thus triggering a signal. We want to find out under what conditions the negative test quantity will have returned to zero.

Let us put the number of successive times at which the positive (negative) test quantity has been different from zero, up to and including the rth observation, equal to $m(n)$. (m, n can be zero.)

The test quantities can then be written as

$$S_j = \sum_{i=t}^{t+j-1} (e_i - k^+), \quad s_l = \sum_{i=u}^{u+l-1} (e_i + k^-)$$

and

$$0 < S_j < h^+ \quad (1 \leqslant j \leqslant m)$$

$$0 > s_l > -h^- \quad (1 \leqslant l \leqslant n), \text{ with } S_0 = s_0 = 0.$$

t is the starting observation of accumulation of the positive test quantity and u that of the negative test quantity. We have:

$$t + m - 1 = u + n - 1 = r.$$

As the $(r+1)$th observation causes the positive test quantity to exceed h^+, we have:

74

$$S_{m+1} = S_m + e_{r+1} - k^+ \geqslant h^+.$$

Hence
$$e_{r+1} \geqslant h^+ + k^+ - S_m. \tag{A2}$$

The negative test quantity now becomes (using (A2))

$$s_{n+1} = s_n + e_{r+1} + k^- \geqslant s_n - S_m + h^+ + k^+ + k^-. \tag{A3}$$

To satisfy condition (A1), the right-hand member of (A3) must always be non-negative. The difference $s_n - S_m$ can be worked out for the three different cases which can occur, depending on which sum started accumulating first. We obtain

$$s_n - S_m + h^+ + k^+ + k^-$$

$$= \begin{cases} h^+ + (m+1)(k^+ + k^-) + s_{t-u} & (t > u) \\ h^+ + (m+1)(k^+ + k^-) = h^+ + (n+1)(k^+ + k^-) & (t = u) \\ h^+ + (n+1)(k^+ + k^-) - S_{u-t}. & (t < u) \end{cases} \tag{A4}$$

The second and third cases are always non-negative, no matter what the cusum parameters are. For the first case we have

$$h^+ + (m+1)(k^+ + k^-) + s_{t-u} \geqslant h^+ + k^+ + k^- + s_{t-u} > h^+ - h^- + k^+ + k^-.$$

Hence the negative test will always have returned to zero if

$$h^+ - h^- + k^+ + k^- \geqslant 0$$

or
$$k^+ + k^- \geqslant h^- - h^+. \tag{A5}$$

Conversely, if the negative test signals, the positive test will always have returned to zero if

$$k^+ + k^- \geqslant h^+ - h^-. \tag{A6}$$

Hence a sufficient condition for (A1) to be true is

$$k^+ + k^- \geqslant |h^+ - h^-|. \tag{A7}$$

Corollary 1: By substituting relations (10) into (A7), formulae (23) and (24) of page 21 follow. Note that (24) is the above-mentioned result by Kemp (1961), but with the strict inequality replaced by the weaker one.

Corollary 2: An odd consequence of the way in which $s_n - S_m$ builds up is that — for a symmetrical test — both test quantities cannot be each different from zero and yet produce no signal for longer than $h/2k$ observations.

We shall now prove that (A7) is also a necessary condition for statement (A1) to be true. For let us suppose that

$$k^+ + k^- < |h^+ - h^-|$$

and, without loss of generality, take the case that

$$k^+ + k^- < h^+ - h^-,$$

then it is always possible to construct a series of e_i, such that

$$S_m = h^+ - \epsilon \quad \left(0 < \epsilon < \tfrac{1}{2}\{h^+ - h^- - k^+ - k^-\}\right)$$

$$s_n = 0.$$

If the next observation is

$$e_{r+1} = -h^- - k^- - \eta \quad \left(0 < \eta < \tfrac{1}{2}\{h^+ - h^- - k^+ - k^-\}\right)$$

then a signal will be triggered at the lower boundary, whereas

$$S_{m+1} = h^+ - h^- - k^- - k^+ - \epsilon - \eta > 0.$$

REFERENCES

References are shown in the standard way, with one exception. In front of each reference a capital A, B, C, D, E indicates the class of papers into which the paper falls. Of course, there are cases in which the classification is ambiguous; in those cases we have chosen the class considered to be most suitable. The classes are:

A: Classical control charts

B: Cumulative sum charts or tests — general.

C: Cumulative sum tests with particular reference to solution of the integral equations.

D: Adaptive optimization, etc.

E: Routine prediction and tracking signals.

C   Anscombe, F.J. and Page, E.S. (1954), "Sequential Tests for Binomial and Exponential Populations." *Biometrika*, **41**, 252 – 253.

B   Armitage, P. (1950), "Sequential Analysis with more than Two Alternative Hypotheses and its Relation to Discriminant Analysis." *J.R.S.S.*, B, **12**, 137 – 144.

A   Barnard, G.A. (1954), "Sampling Inspection and Statistical Decisions." *J.R.S.S.*, B, **16**, 151 – 174.

B   Barnard, G.A. (1959), "Control Charts and Stochastic Processes." *J.R.S.S.*, B, **21**, 239 – 270.

C   Barraclough, E.D. and Page, E.S. (1959), "Tables for Wald Tests for the Mean of a Normal Distribution." *Biometrika*, **46**, 169 – 177.

    Bartlett, M.S. and Kendall, D.G. (1946), "The Statistical Analysis of Variance-Heterogeneity and the Logarithmic Transformation." Suppl. *J.R.S.S.*, **8**, 128 – 138.

D   Box, G.E.P. and Jenkins, G.M. (1962), "Some Statistical Aspects of Adaptive Optimization and Control." *J.R.S.S.*, B, **24**, 297 – 343.

D   Box, G.E.P. and Jenkins, G.M. (1963), "Further Contributions to Adaptive Quality Control; Simultaneous Estimation of Dynamics: Non-zero costs." University of Wisconsin, Technical Report No.19 and Bulletin of the I.S.I., 34th session, Ottawa.

E   Brown, R.G. (1959), *Statistical Forecasting for Inventory Control*. McGraw-Hill, New York.

E   Brown, R.G. and Meyer, R.F. (1961), "The Fundamental Theorem of Exponential Smoothing." *J. Operat. Res. Soc. Amer.*, **9**, 673 – 686.

E   Brown, R.G. (1963), *Smoothing, Forecasting and Prediction of Discrete Time Series* (especially Chapter 5). Prentice Hall, New York.

C   Courant, R. and Hilbert, D. (1953, English edition), *Methods of Mathematical Physics*, vol. 1. Interscience Publishers, New York.

E   Cox, D.R. (1961), "Prediction by Exponentially Weighted Moving Averages and Related Methods." *J.R.S.S.*, B, **23**, 414 – 422.

78

C   Cox, D.R. (1962), *Renewal Theory*. Methuen & Co Ltd., London, and
      Wiley & Sons, New York.

B   Davies, O.L. (ed.)(1963), *The Design and Analysis of Industrial Experi-
      ments*. Oliver & Boyd, London and Edinburgh.

E   van Dobben de Bruyn, C.S. (1963), "Smoothing Sales Data by Progressive
      Correction". *Statist. Neerl.*, **17**, 243 – 265.

D   van Dobben de Bruyn, C.S. and Muijen, A.R.W. (1964), "The Effect of In-
      formation Delays in a Production Control System." *Int.J.Prod.
      Res.*, **3**, 167 – 175.

E   van Dobben de Bruyn, C.S. (1964), "Prediction by Progressive Correction."
      J.R.S.S., **B, 26**, 113 – 122.

E   van Dobben de Bruyn, C.S. (1967), "The Interplay of Tracking Signals and
      Adaptive Predictors." *The Statistician*, **17**, No.3, 1 – 10.

A   Dodge, H.F. and Romig, H.G. (1944), *Sampling Inspection Tables*. Wiley &
      Sons, New York.

A   Dudding, B.P. and Jennett, W.J. (1942), "Quality Control Charts."
      British Standard 600R. London, British Standards Institution.

B   Ewan, W.D. and Kemp, K.W. (1960), "Sampling Inspection of Continuous
      Processes with no Autocorrelation between Successive Results."
      Biometrika, **47**, 363 – 380.

B   Ewan, W.D. (1963), "When and how to use Cusum Charts." *Technometrics*,
      5, 1 – 22.

C   Feller, W. and Orey, S., (1960), "A Renewal Theorem." *J. Math. and Mech.*,
      10, 619 – 624.

C   Feller, W. (1966), *An Introduction to Probability Theory and its Applica-
      tions*, vol. 2. Wiley & Sons, New York.

A   Freund, R.A. (1962), "Graphical Process Control." *Ind. Qual. Control*, **18**,
      15 – 21.

B   Goldsmith, P.L. and Whitfield, H. (1961), "Average Run Lengths in Cumu-
      lative Sum Quality Control Schemes." *Technometrics*, **3**, 11 – 20.

A   Grant, E.L. (1946), *Statistical Quality Control*. McGraw-Hill, New York
      and London.

E   Grünwald, H. (1965), "The Correlation Theory for Stationary Stochastic
      Processes applied to Exponential Smoothing." *Statist. Neerl.*,
      19, 129 – 138.

E   Harrison, P.J. and Davies, O.L. (1964), "The Use of Cumulative Sum
      (CUSUM) Techniques for the Control of Routine Forecasts of
      Product Demand." *J. Operat. Res. Soc. Amer.*, **12**, 325 – 333.

E   Harrison, P.J. and Scott, F.A. (1965), "A Development System for use in
      Short-term Forecasting Investigations." Paper read at Opera-
      tional Res. Soc. Conference, September 1965, Shrivenham.

E   Harter, H.L. (1964), "New Tables of the Incomplete Gamma-Function
      Ratio and of the Percentage Points of the Chi-Square and Beta
      Distributions." U.S. Government Printing Office, Washington D.C.

C   Hastings, C. (1955), *Approximations for Digital Computers*. Princeton Uni-
      versity Press, Princeton, N.J. and Oxford University Press,
      London.

C   H.M.S.O. = Her Majesty's Stationery Office (1955), *Interpolation and Allied
      Tables*. Prepared by order of the Lords Commissioners of the
      Admiralty by H.M. Nautical Almanac Office. London.

E   Holt, C.C. (1957), "Forecasting Seasonals and Trends by Exponentially Weighted Moving Averages." Pittsburg, Penn.: Carnegie Institute of Technology.

C   Jöhnk, M.D. (1964), "Erzeugung von Betaverteilten und Gammaverteilten Zufallszahlen." *Metrika*, **8**, 5 – 15.

B   Johnson, N.L. (1961), "A simple Theoretical Approach to Cumulative Sum Control Charts." *J. Amer. Statist. Ass.*, **56**, 835 – 840.

B   Johnson, N.L. and Leone, F.C. (1962), "Cumulative Sum Control Charts: Mathematical Principles applied to their Construction and Use." *Ind. Qual. Control*, June, 15 – 21; July, 29 – 36; August, 22 – 28.

B   Johnson, N.L. (1963), "Cumulative Sum Charts for Folded Normal Distribution." *Technometrics*, **5**, 451 – 458.

C   Keilson, J. (1965), *Green's Function Methods in Probability Theory.* Charles Griffin & Co Ltd., London.

C   Kemp, K.W. (1958), "Formulae for Calculating the Operating Characteristic and the Average Sample Number of some Sequential Tests." *J.R.S.S.*, **B, 20**, 379 – 386.

B   Kemp, K.W. (1961), "The Average Run Length of the Cumulative Sum Chart when a V-mask is used." *J.R.S.S.*, **B, 23**, 149 – 153.

B   Kemp, K.W. (1967a), "Formal Expressions which can be used for the Determination of the Operating Characteristic and Average Sample Number of a Simple Sequential Test." *J.R.S.S.*, **B, 29**, 248 – 262.

B   Kemp, K.W. (1967b), "A Simple Procedure for determining Upper and Lower Limits for the Average Sample Run Length of a Cumulative Sum Scheme." *J.R.S.S.*, **B, 29**, 263 – 265.

C   Kemperman, J.H.B. (1963), "A Wiener-Hopf Type Method for a General Random Walk with a Two-sided Boundary." *Ann. Math. Statist.*, **34**, 1168 – 1193.

C   Latter, R. (1958), "Approximate Solutions for a Class of Integral Equations." *Quart. Appl. Math.*, **16**, 21 – 31.

C   Lovitt, V.W. (1924), *Linear Integral Equations.* McGraw-Hill, New York, reprinted 1950 as Dover Publication S 176, New York.

Lyle, P. (1954) "The Construction of Nomograms for use in Statistics." *Appl. Statist.*, June 1954, 116 – 125.

E   Meyer, R.F. (1963), "The Optimality of Exponential Smoothing." (In discussion section on Forecasting at Third Int. Conf. on Operat. Res. (IFORS), Oslo, July 1963.)

C   Mikhlin, S.G. (1957), *Integral Equations in Mechanics, Physics and Technology.* Pergamon Press.

E   Muth, J.F. (1960), "Optimal Properties of Exponentially Weighted Forecasts." *J. Amer. Statist. Ass.*, **55**, 299 – 306.

E   Nerlove, M. (1963), "On the Optimality of Adaptive Forecasting." Paper presented at the 2nd European Congress of the I.M.S., Copenhagen, July 1963.

C   Noble, B. (1958) *Methods based on the Wiener-Hopf Technique.* International Series of Monographs in Pure and Applied Mathematics, Pergamon Press.

A   Olds, E.G. (1961), "Power Characteristics of Control Charts." *Ind. Qual. Control*, **17**, 4 – 10.

B   Page, E.S. (1954a), "Continuous Inspection Schemes." *Biometrika*, **41**, 100 − 115.

C   Page, E.S. (1954b), "An Improvement to Wald's Approximation for some Properties of Sequential Tests." *J.R.S.S.*, **B, 16**, 136 − 139.

B   Page, E.S. (1957), "On Problems in which a Change in a Parameter occurs at an Unknown Point." *Biometrika*, **44**, 248 − 252.

B   Page, E.S. (1961), "Cumulative Sum Charts." *Technometrics*, **3**, 1 − 9.

A   Page, E.S. (1962a), "A Modified Control Chart with Warning Lines." *Biometrika*, **49**, 171 − 175.

B   Page, E.S. (1962b), "Cumulative Sum Schemes using Gauging." *Technometrics*, **4**, 97 − 109.

B   Page, E.S. (1963), "Controlling the Standard Deviation by Cusums and Warning Lines." *Technometrics*, **5**, 307 − 315.

B   Roberts, S.W. (1966), "A Comparison of Some Control Chart Procedures." *Technometrics*, **8**, 411 − 430.

A   Shewhart, W.A. (1931), *Economic Control of Quality of Manufactured Product*. Van Nostrand, New York.

C   Spitzer, F. (1957), "The Wiener-Hopf Equation whose Kernel is a Probability Density." *Duke Math.J.*, **24**, 327 − 343.

C   Titchmarsh, E.C. (1937), *Introduction to the Theory of Fourier Integrals*. Clarendon Press, Oxford.

C   Tocher, K.D. (1963), *The Art of Simulation*. English Universities Press Ltd., London (Electrical Engineering Series).

B   Torrens-Ibern, J. (1965), "Les methodes statistiques de contrôle dans les processus industriels continues." *Rev.Statist.Appliquée*, **13**, 65 − 93.

E   Trigg, D.W. (1964), "Monitoring a Forecasting System." *Operat.Res.Quart.*, **15**, 271 − 274.

B   Truax, H.M. (1961), "Cumulative Sum Charts and their Application to the Chemical Industry." *Ind.Qual.Contr.*, **17**, 18 − 25.

B   Wald, A. (1947), *Sequential Analysis*. John Wiley & Sons, New York.

E   Ward, D.H. (1964), "Comparison of Different Systems of Exponentially Weighted Prediction." *The Statistician*, **13**, 173 − 191.

C   Wiener, N. and Hopf, E. (1931), "Ueber eine Klasse singulärer Integralgleichungen." *S.B.Preuss.Akad.Wiss.*, 1931, 696 − 706.

B   Woodward, R.H. and Goldsmith, P.L. (1964), *Cumulative Sum Techniques.* I.C.I. Monograph No. 3, Oliver & Boyd, Edinburgh.

E   Yaglom, A.M. (1955), "Correlation Theory of Processes with Stationary nth Differences." (In Russian) *Mat.Sbornik*, **37**, 141 − 196.